DATE			

FIELDS WITHOUT DREAMS

Other Books By Victor Davis Hanson

Warfare and Agriculture in Classical Greece

The Western Way of War: Infantry Battle in Classical Greece

Hoplites: The Classical Greek Battle Experience (editor)

The Other Greeks: The Family Farm and the Agrarian Roots of Western Civilization

FIELDS WITHOUT DREAMS

—Defending the Agrarian Idea—

VICTOR DAVIS HANSON

THE FREE PRESS

New York London Toronto Sydney Tokyo Singapore

The Free Press
A Division of Simon & Schuster Inc.
1230 Avenue of the Americas
New York, N.Y. 10020

Printed in the United States of America

printing number

2 3 4 5 6 7 8 9 10

Text design by Carla Bolte

Library of Congress Cataloging-in-Publication Data

Hanson, Victor Davis.
Fields without dreams: defending the agrarian idea / Victor Davis Hanson.
p. cm.
ISBN 0-684-82299-7
1. Family farms—United States. 2. Agriculture—Economic aspects—United States. I. Title.
HD1476.U5H28 1996
338.1'6—dc20 95-40875
CIP

To the memory of

Frank, Fannie, and Al,

and Rees, Georgia, Lila, Lucy, and dear Pauline,

and to all the past Hanson and Davis yeomen whose land we

—Alfred, Rees, Nels, Maren, Victor, and William—

also tried to preserve and so to pass on.

CONTENTS

PREFACE

Much is being written now about family farming in America. The more farms pass away, the more publications appear. Certain tenets of this growing genre are easily identifiable in these books and they remain mostly unquestioned: Family farmers are noble, but vanishing stewards of ancestral ground—in contrast to a grasping agribusiness, whose singular profit mentality has turned nearly all of agriculture into little more than an industry, where chemicals go into the soil and profits alone emerge.

I have no objection per se to that stereotyped indictment of corporate farming that has now become American farming. Indeed, many of those similar conclusions are probably found here in this book. Where I differ, however, from most of what is written about the American family farm is in tone and historical perspective. Most recent exposés and memoirs belong to the lyrical, romantic—and often naïve—tradition that is certainly neither new nor necessarily American, but was inaugurated long ago by the Roman poet Virgil. His *Georgics* ("farming things") romanticized the harmony and community of the countryside, in implicit contrast to the bustle and impersonality of urban life, emphasizing the nobility of the farmer and the natural beauty of his craft. His is a picture, then, that is comforting and easy to embrace.

Most recent writers on farming belong to this one-dimensional agrarian genre that grew out of the *Georgics*, and are oblivious to the fact that there is also another—much older, bleaker, and mostly unknown tradition—begun seven centuries earlier by the Greek poet Hesiod. His *Works and Days* as a more melancholy, more angry account of the necessary pain and sacrifice needed to survive on the land, did not find, and has not found, Virgil's popular audience. No wonder: Hesiod's soil is not kind, but unforgiving, and so must be mastered if it is not to master the farmer himself. Those who follow that regimen, in Hesiod's eyes, are *not* necessarily sympathetic or sensitive individuals, but often appear as more unattractive folk as the cost of their wager to live off the land. So I agree more with Hesiod that farmers have always been distasteful, independent, and of a different sort locked in perpetual struggle with the "bribe-swallowers" in town, the princes who profit from, but do not partake of, an agrarian community. Hesiod's poem of agriculture is not merely a didactic treatise on agricultural technique, nor, like Virgil's, a paean to farming, but a case study in the agrarian profile, and hence an alternative paradigm of values of society itself. So Hesiod is often a voice we do not wish to hear of, much less to follow.

With Hesiod's world begins the entire notion of agrarianism that was soon to become the foundation of the Greek city-state, and later to be enshrined in the West as the exemplar of a democratic society: a culture of small, independent yeomen on the land, who make their own laws, fight their own battles, and create a community of tough like-minded individuals. Whatever one thinks now of Western culture, he should at least recognize that its foundations—economic, social, political, and military—originated in the countryside, and so must direct either his anger or praise toward those few descendants who still farm as in the beginning.

From Hesiod we learn that raising food from the soil is not a clean experience; living by that toil is not necessarily always moral;

passing that work on is often nearly impossible—and that entire experience is not unique to America at the millennium. We are at the end of a historical cycle in America, which has occurred many times in the history of civilization. Agrarianism emerges on the detritus of a failed complexity; it creates a stability that leads to affluence and greater freedom; that bounty invariably leads to a rejection of a now boring and unimaginative community of small and blinkered farmers. Any book about farming must now not be romantic nor naive, but brutally honest: The American yeoman is doomed; his end is part of an evolution of long duration; and so for historical purposes his last generation provides a unique view of a world—a superior world I will argue—that is to be no more.

The grapes of wrath have been over for nearly a century now. American farmers no longer struggle for their very existence. The fight has ceased to be one of shelter, sustenance, livelihood, or of salvation itself. Any yeoman can easily become a well-fed clerk, truck driver, or salesman, once he is moved off the land. Nor is farming in America any longer a field of dreams, a romantic hideaway of the mind where folks can return to childhood innocence and the security of clapboard houses, cornfields, and baseball. That stage of agrarianism in the American consciousness I think should also now pass on, because the quaint family farmstead, the focus for such fantasy, is itself becoming a caricature, not a reality, in the here and now.

We farmers are neither the Joads, who starve in the fields of another, nor the touched, who hear voices and plow up a few acres for baseball fields. Rather, I mean by this title, *Fields Without Dreams*, that we have now entered the third wave of American agrarianism beyond the first and largely existential, and that second that was mostly romantic. We are now in the third stage of a future that has no future, an agrarian Armageddon at the millennium where the family farm itself—both as a way of life and a reassuring image of the mind—will be obliterated. And its credo, the agrarian idea, is what I intend to present and defend here in this postmortem.

Again, others can write of the harm of chemicals and corporations, of the beauty and security of their farms, but I am interested in the agrarian and his relationship to the larger world, ancient and modern. Better to confess in tones of defiance that we are not always losing better people, but simply *different* people whose like we are not about to see again in our lifetime. Nor does defending the agrarian idea mean offering a blueprint for a humane agriculture—although throughout this book I offer positive suggestions for needed changes in the countryside—but rather leaving behind a brazen warning: This bothersome, queer oddball who is disappearing has been for twenty-five hundred years the critical counter voice to a material and uniform culture that at its basis is neither democratic nor egalitarian.

Thus this book throughout asks the same question about the consequences of his absence: *Where* will be the often unpleasant individual, the veteran of a continual struggle with nature, the now cultural dissident, who will choose still to go it alone in order to protect his old notion of a community, who will by nature have distrust for authoritarianism, large bureaucracy, and urban consensus? Is there another besides the ugly agrarian whose voice says no to popular tastes, no to the culture of the suburb, no to the urban enclave, no to the gated estate? What other profession is there now in this country where the individual fights alone against nature, lives where he works, invests hourly for the future, never for the mere present, succeeds or fails largely on the degree of his *own* intellect, physical strength, bodily endurance, and sheer nerve?

Finally if there is no future in family farming in America, why does the last generation continue to work these fields where there are no dreams? Yeomen are not naive. They are not all blinkered simpletons who believe in "just one more year" and things will yet ameliorate. Nor do they think that they alone ensure us of food that is cheap, safe, and nutritious. I know not one who claims that defending the agrarian idea means his presence is vital to the continuation of American democracy. All that is the abstract rationale of their more erudite defenders. Farmers themselves, I think, have no illusions that they are nearing the last harvest. No, they continue in their fields without dreams and defend their agrarian idea for two very simple, concrete—and, in some sense, selfish—reasons.

First, no one wishes to be the failed link in a great chain of succession. For all those born and raised on the farm, the horse and plough, the Great Depression, World War II, crop failure, and bank collapse are not merely the familiar litany of the agrarian lament. They are instead yardsticks to measure one's self against those who came before. That agrarian template puts your own debacle with the bank and broker into its proper perspective as part of the age-old fight between those who grow and those who sell. To leave now, cash out, and be cast adrift—even when prudence and fairness to those in the present demand precisely that path—is to admit inferiority before the judges omnipresent with the private memory. From childhood those voices that every farmer carries in his brain whisper, must whisper, "We survived, where you cannot; we took nothing, where you have taken everything." To sign one more time on the bank's dotted line is suicide, but not to sign is a betrayal of those on the other side.

Second, the incentive to keep farming when one should not is not entirely quixotic. Farmers chauvinistically prefer, I think, that their progeny resemble their parents and grandparents rather than their peers in contemporary America. Even as they realized that their farms will not last and not be passed on, so all the more adamantly they embrace their fields now as a steady nursery in lieu of a real livelihood. "At least the kids had somewhere to grow up," I've known them to say, as if housing and communities were scarce elsewhere, as if, as Aeschylus said, learning arrives only through ordeal and trial.

In the farmers' mind, given their pessimistic views of urban America in the late twentieth century, there is no other avenue to ensure their children a moral future, save by putting them on the land and putting them to work. Only that way, for the rest of their lives will they have a house, not houses, have a job, not jobs, have a town, not towns, have a family, not families, know where they were born and where they will die. On the farm they will have a bothersome cargo, it is true, that sets them apart, restricts their vision, clouds their liberality, ensures they will not be players on the American scene. If ours was a regimented society, as in the past where tradition was confining and custom arbitrary, the agrarian could perhaps be too stern and so only add to an oppressive conformity. But as the farmer knows now, that is *not* currently our problem in this country.

There are enough artists, writers, provocateurs, purveyors of fad, cant and chaos, enough sophisticated and educated to suggest what we see and hear is not what it seems, enough shiftless and transient who reinvent themselves with each move. But at the millennium there are far too few of the other in America, the traditional and time-honored brakemen on an affluent, leisured, and rootless society. There is a reason why farmers defiantly, brazenly, want their children raised differently, want them to subtract from, not add to, the current American madness, want them, I suppose, to be like themselves: to have fields without dreams, rather than dreams without fields.

INTRODUCTION

I.

The following account is what I either saw or heard between 1981 and 1993 on or near our small farm in the San Joaquin Valley of California. Most of the experiences took place between 1981 and 1985.

At the time these events transpired, I had no idea they were anything but transitory and solitary phenomena, without much significance other than to my immediate family. I did not envision that we few, ignorant and isolated, were one with thousands of vanishing farm families; that we, in California of all places, were kindred with those in Nebraska, Iowa, and Georgia. Nor did I suspect that 1983 would mark the beginning of an abrupt nationwide resumption of decades of ongoing agricultural depression, one that saw the disappearance of about two thousand family farms a week through much of the decade.

There has not been a single month since the early 1950s when the number of genuine American farms has increased or even stayed the same—a fact often overlooked by economists who say the erosion of family farms has been only "gradual." Like 5 percent losses on B-17 missions over Europe, that steady reduction is not sustainable; it finally takes your whole squadron, and so—even by the most jaded planner—is deemed unacceptable. On October 8, 1993, for example, it was reported that the Census Department had ceased counting the number of American farmers. Those still making a living by working and living on their own land were dubbed "statistically insignificant." And it is not quite over yet.

Although the farming of trees and vines is a special subspecies of agriculture, what follows is not the idiosyncratic tale of a few thousand raisin and plum growers in central California. It is but a ripple of a much larger concussion that has changed the face of American agriculture in the last decade. True, from the freeway American farms look about the same. The food is still plentiful in the grocery store. Genuine agrarians remain to be found at the ubiquitous farmers' markets, the agricultural equivalent of theme parks and petting zoos. But on closer examination, an entire generation of farmers—families whose sole income derived from raising food on ancestral ground—vanished, and the consequence of that loss has yet to be determined. Most farm income has either fallen or been stagnant for more than a decade as expenses have more than doubled—an incendiary situation for any other profession, whether schoolteacher, banker, or insurance agent.

For the agricultural economist and financier, I have learned, it makes little difference whether fifty thousand or five million Americans own farms, whether the owner has title to one hundred or fifty thousand acres, whether the men and women on tractors live where they work or, like the majority, are anonymous commuters. But I leave the rebuttal to that argument to farm activists and the more romantic (consult the plethora of books in the HD section of the university library), who warn, often shrilly, of environmental catastrophe, salination, species deprivation, soil erosion, and eventual food shortages—the apocalypse to come under the corporations whose prime directive in growing food is sure profit in the here and now, no more.

Again, as stated, I am more worried about the cultural and historical ramifications of the elimination of an entire genus of American, albeit an American intrinsically no better or worse than the factory worker, secretary, or high school teacher. The farmer is perhaps not inherently noble, law abiding, or even ethical. But he is different, vastly different, from almost all other types of citizens, whose methods of wage labor, whose house, whose values and material outlook, are now all roughly similar. He is distinct because he is shaped solely by the growth of plants, his success dependent on his degree of skill in coercing food from the earth, his very family predicated on their own utility in aiding that singular cause. Physical labor, the uncertainty of recompense, and solitude are the welcomed, not the despised, wages of agriculture, the "best tester of good and bad men" as the Greeks said. We know the aggregate effect of such yeomen on the commonwealth: For most of the seventeenth, eighteenth, and nineteenth centuries they were, again for good or evil, the majority of the North American population—95 percent at the founding of the republic—and so their valuation depends on your own estimation of America itself.

Even in the first third of the twentieth century, their numbers were to be counted in the millions. But their complete absence from the American landscape will be the great legacy of the present generation; agrarians—families whose sole support, whose only occupation, is growing food—at the end of the millennium may well have vanished. It is my simple contention, supported solely by instinct and supposition, that the entire cargo of our current unhappiness—materialism, crime, spiritual emptiness—is in inverse proportion to the number of people who are both rural and agrarian.

The life of a farmer is physical. It is concrete. As a part of nature and the unfathomable processes of the plant kingdom, it is also both spiritual and mystical. It is a struggle with a purpose, where self and family have clear roles prescribed by the wisdom of the ages. As both Xenophon and Aristotle saw, there is little exploitation of one's fellow man in agriculture, "the mother of us all," where the struggle against nature makes one bigger, not smaller. The larger the number of people engaged in that activity, the greater the brake on the rest of us. Agrarians need not be the majority of the population—although Aristotle felt their overwhelming presence alone created a stable citizenry—but they do need to be present in large enough numbers to be heard, to offer a shake of the head, when asked, that sends a tremble and a quiver to the majority to halt and desist.

II.

If there is a unity in these brief remembrances, I think it derives from three themes.

(1) The production of food is not in immediate danger. The short-term status of America's soils and fertility is assured. But family farming—for a variety of complex economic, political, and cultural reasons—has for much of this century been in decline. In the past decade this erosion has accelerated to such a degree as to suggest that most true agrarians will be gone before the millennium is over. Therefore, for purely historical purposes, I felt a need to chronicle this world as it disappears without abstract analysis or formal research. Antiquarianism has had a noble pedigree in the West, so I make no apology for the recollection of farming events as I saw them. What went on during this last generation of family farming should not be entirely lost to the next century. Indeed, my hope—entirely unfounded if not preposterous—is that the historical agricultural cycle, the recurring destruction and resurrection of homesteaders, has not been wholly broken by technology and science. Might not, a century or two hence, agrarians yet again reemerge from the ashes of contemporary latifundia? If so, this book may also offer some small help and solace to struggling farmers not yet born.

(2) In an age when most of us are urban and conditioned by the behavioral and social sciences—our speech, thought, education, law, and religion now guided by psychology, anthropology, sociology, linguistics, and political science—it is critical to understand that this

assumption was neither always the case nor often thought to be a good thing. There is a tiny fraction of the American population whose code of conduct and outlook is still predicated on very different premises, on exactly the opposite idea of social construction, on criteria more natural yet distinct from the supporting apparatus of religion and abstract science. Staying in one place, working with one's hands, challenging nature through group struggle, passing on something better than what one received—this creed can manifest itself in peculiar ideas about the rest of us.

People to the farmer's eye, I shall suggest, are good or no good, lazy or industrious, kind or cruel. Period. Their tasks are either left incomplete or accomplished. Men—all men, regardless of race, age, or environment—either kill or do not slay, steal or do not loot. To the farmer—who either produces or fails to produce food, who either masters nature or is mastered by it—much of what passes in political America—Democratic and Republican alike—as legitimate, as just and moral, is found wanting. I offer this alternative view of our culture because I worry that when the agrarian yardstick has vanished completely, there will be no bridle on the present absurdity of random violence, growing illiteracy, and spiritual desolation, no one left to tell us how silly it is all becoming.

True, cultural conservatives abound. Many are articulate critics of what is wrong with this country. But too often their conservatism is simply financially based and so without deep roots; little more than the worry over a reckless government expropriating their hard-won dollars to waste on failed entitlements. Or sometimes it is counterfeit, the darker reaction engendered by race, the discomfort with the changing face of an America that is to be a third to a half nonwhite and not comfortable with the culture of the West. Others still—given the general aging of the population—are simply romantic reactionaries. They nostalgically contrast the general tranquillity of their earlier years with the crime-ridden and unsafe America of the 1980s and 1990s. So it is increasingly difficult anymore to find a natural bedrock conservative, someone whose unease with present fashion and trend—whether it be multiculturalism or leveraged buyouts—is rooted in a lifetime's observance of the growth and decay of plants; in the idea that a house, like a barn or shed that protects from the elements, need not be torn down; in the notion that men, like trees and vines, wear, age, and die and so have no need of cosmetic restoration, no worry over ample bellies and lost hair.

Even frost, drought, infestation, and flood are explicable and perhaps for some good purpose, the farmer learns. The agrarian, however, becomes most acutely mystified over human behavior that now seems to have no purpose, that is not even rooted in the age-old cycle of acquisition and consumption. His political views—nearly exclusively Republican—are incidental to his conservatism. He expresses more a knee-jerk and blanket distaste for fashion, affluence, and leisure—the current cradle of criminality for the urban rich and poor alike. In other words, his own world is an island of absolutes in a sea of relativism.

(3) Finally, like old Ajax in Sophocles' masterful play (his greatest play), the vanishing agrarian has a nobility apart from any question of good or evil. I am not sure, as the reader will quickly learn, whether the yeoman is altogether moral or amoral. I am convinced that he possesses dignity because he believes in something unconnected with money, status, and power, something that is fated and unwanted on the American landscape. This choice between an identity and the rewards of the here and now is known to most farmers. Yet most agrarians would rather pass away than change to join the rest of us in the comfortable pond of pelf and security. However silly and corny that all seems, it still is rare to find in modern America anyone who will pay in currency and health to have a lifetime free of nodding "yes." In what follows, I suggest why it is that agrarians do not change and so surely will continue to disappear.

Remember also that the end of family farming is an insidious process. It is not a black-and-white morality tale—at least not entirely—where the power of evil vanquishes the last defenders of the true faith. Good men of four generations' lineage find it noble to sell out and use the money to send children to college, while the more selfish, duplicitous, and unjust can cajole and cheat to remain on the land. Neighbors grow and grow—often through hard work and financial acumen—gobbling up the failed adjacent, until through equity and subsidy they too enter the world of the corporate mogul, and so find salvation through the wreckage of their brethren.

At many a Christmas dinner in America, a nephew who has drained the financial reserves of an entire family to abide on family land sits across from his sibling, a dutiful officer for the Land Investment Company whose stock and trade is the acquisition of failed farms for the profits of distant investors—someone who would never ask another to subsidize the upbringing of his own offspring. At the end of the table are the in-laws from Los Angeles, two of four or five million in this country who are unhappy at urban work and so like the abstract idea of farming, its solitude and natural setting. Perhaps they have three or four million in deferred income that they might place on the table for a shot at replacing the failed yeoman's slot in the vanishing grid. Indeed, this process of agrarian roulette ensures that land is kept in production, as the capital-laden romantics rush in and lose money for half a decade or so in American agriculture, so strong has become the silhouette of the doughty agrarian conjured in their brains.

What is common to all this, then, is not merely the moral and the base, corporation versus family, urban against rural. Rather, it is a fascinating leap into a completely different world, the departure of the family farm in the 1980s every bit as hasty and abrupt as the onslaught of the information age of cellular phones and E-mail. You see, we are now ending a very old idea in this country—a belief that goes back to the Greeks of the *polis*—that a family inherited land, grew food, and was rewarded with a life that fed and clothed children and that such agrarianism had value to all beyond the confines of the farm. As the notion passes that such people were the foundry of the country—its values, its militia, its very resilience—it is not altogether clear where, or if, they are to be replaced and what is to be left in their wake.

What follows, then, are the accounts of this agrarian madness of 1983 and after, how it challenged but ultimately destroyed many of the people I knew. I say "knew" because even the few healthy survivors I still see are no longer those whom I knew. Life appears to go on, but wounds of a decade of that magnitude eventually reach the vital organs, fed and nourished by an evil triad of bitterness, frustration, and remorse. In the little community of vines around our farm, the tally was about average: Kaldarian dead; Kapoor dead; Heine and Barzagus heart attacks; uncle suicide through self-inflicted shotgun wound; mother brain tumor; a few other kindred laid up sick or gone crazy—just about the average toll for four or five square miles of Thompson grape vineyard. Each of these following dramas derived, I think, in some way, from 1983, from that sudden, abrupt end to the entire notion of families making a living from farming raisins. The people who fell sick, died, went alcoholic, chose suicide, or simply gave up and moved away—I see now for the first time, a decade later, that they all had the origins of their ruin in that most awful year, 1983.

III.

Unfortunately, when most of these events first took place, I never foresaw that I would ever again write anything beyond an esoteric and little-read monograph on ancient Greek warfare and agriculture. I was a farmer, not a professor, much less a writer of anything of value. Consequently, I took no written notes of occurrences as I farmed. Now, more than a decade after 1983, I am forced to rely largely on memory and what unexpected information I recently found collected in the files of my late mother. In many cases, where memory fails and there are no documents, bank records, or newspaper clippings, I sought confirmation from friends and family members, themselves, for better or worse, many of the principals involved in these accounts.

At key places in the text, I have inserted direct speech in quotation marks, the most memorable of the thousands of conversations held during the last fifteen years. Farmers love to talk about their impending doom and the "heroic" measures they must take to seek their salvation. While the exact wording of each quoted conversation may vary slightly from what was actually uttered, I believe nearly all are accurate representations of what was actually said.

I have given *all* the characters, businesses, some crop varieties, and most of the locations in these accounts pseudonyms; the names of family members are altered as well. In addition, in a very few instances some of the descriptions and incidents themselves were changed, elaborated, or disguised so as not to identify those portrayed. You see, there is no small measure of self-interest here as well: I continue to live and work and will die among the people who appear in the following pages. Finally, I have avoided citation of secondary work on the destruction of American family farming during the recent decade. That more formal academic account can be found in a variety of good books and journal articles, and I have written about it at some length in a recent history, *The Other Greeks: The Family Farm and the Agrarian Roots of Western Civilization* (The Free Press, New York, 1995); material from four pages of that book have been freely drawn on here. I wish to thank Bruce Thornton, John Heath, Cara Hanson, and Glen Hartley for reading an early draft of this memoir, and Robert Benard, Adam Bellow, and Mitch Horowitz for editorial suggestions and changes.

—VDH
Hanson-Nielsen Farm
Selma, California

CHAPTER ONE

IN THE BEGINNING

I.

In the beginning here there was nothing. No Thompson seedless vines, no plums or peaches. The nectarines outside the window did not yet even exist as a species. There was no house. No wells, pumps, or pipelines. No roads, no town, no neighbors. In 1878 there was no gas, propane, or electricity here, no power other than wind, horse, and muscle.

But there was land and water on our Alma farm, 120 acres of scrub brush and a pond in a sea of cheap, worthless land the railroad did not need or want. The grandparents of Rhys Burton, our grandfather, hauled dirt up from the pond and planted. Louisa Anna Burton, Rhys's grandmother, and her kin dragged and planted until the pond was small and the land big, until there were five-, ten-, fifteen-acre blocks and more of grapes, fruit trees, tree-lined alleyways, sheds, barns, and houses replete with cousins, nephews, and siblings with queer names like Schuyler, Orin, Cyrus, Emma Lee, Rees, Stanton, Wendell, Marcellus, and Rollo. They built and planted, married and had children, came and left, got sick and died, and are now known to none except those few of us who can still see their sheds and houses and sometimes match their names with faded photographs in forgotten albums.

Only bits of their history survive through haphazard circumstance. Rhys Burton told me that the crooked fifteen vine rows that curve awkwardly up on the hill were planted by his uncle Orin, who was lame and stiff and so could not quite get the rootings next to the planting wire. He told me about Uncle Orin to remind me of the relationship between the farm and the people who farm; one is but the concrete manifestation of the other. We were not simply to grow food but to change the physical landscape around us and so ourselves be altered in the process. I remember that story daily, for at the turn of the century Orin lived in the tackhouse where ninety-five years later I park the cars each night. But, again, the generation of Orin is shadowy, its history little more than hit-and-miss, a family story scarcely remembered and but occasionally repeated; no one any more has any accurate idea who they were and how they lived. It has been, after all, 120 years since they first came here. They are but anonymous images in a Roman household, the ancient Lares whose pictures we dutifully worship.

So their legacy cannot even be preserved through written or oral history but only by the twenty-one different varieties of orchards on their farm that provide fresh fruit all summer long to those who live on the coast, another sixty acres of vines that feed a small city their yearly consumption of raisins, six sheds, five houses, nearly two miles of alleyways, ten miles of subterranean pipelines, five pumps, four families, ten children, and seventy-seven-year-old Jacinta Guevara, to whom my late mother left a house on the farm twenty years ago.

That unwritten, little-recited, but manifestly visible legacy of the Burtons is, I think, what keeps my brother Logan and cousin Rhys farming. Alone in six generations they have planted, pulled, and built to equal Rhys Burton because, they say, the stakes are high, greater than at any time since the Great Depression. They are hurried and relentless, uprooting, planting, grafting and regrafting, borrowing and too often losing, all for all of us, they say, and for everyone who came before. They still believe that there is hope, that after the shakeout of the last two decades, the bounty of fair commodity prices of the seven rich years will yet return for all who would work the land wisely. And where there is hope, I suppose there is still life.

But most disagree. Local real estate agents, the more successful of whom are the proud harvest of agricultural programs at Fresno State and UC Davis and now the top-echelon salesmen for corporations and insurance agencies, correctly point out that real farm income for trees and vines in the valley is at a fifty-year low and that it is time to get out now with the others before you are thrown out by the bank. Three to four thousand acres of orchards and vineyards are, their models indicate, the bare minimum necessary now—and even so will be successful only with the vertical integration of storage, refrigeration, brokerage, and shipping into one operation. So it is hard all the way around; hard to abandon the ethic of all the dead Burtons and farm big as others now do; hard to lose money just to keep the past alive; harder still to bury the dead for good and cash out.

"Would you rather walk away with a hundred thousand or so each, or dribble your equity out over the next decade and end up fifty and broke? It's your call." Ron McCullum, president of Farm Investments Unlimited, the McCullum Group, and RM Associates, told us that recently, as if it were some mystic disclosure known only to his small cloister. This month he pocketed $32,000 in real estate commissions, or so he implied. It is not logical, but childish, to tell a Ron McCullum that you will not sell out because you think you can still make it, that you will not transfer ownership of your tule pond to someone else who does not know of and does not care for a Louisa Anna or Willem Burton—as if a tiny 120-year cycle means much in the billion-year history of the land.

That pond, still Burton Pond on some county maps, is now nearly forty feet deep and dry—deep, for its earth was always needed elsewhere; dry, because the Valley must now water the lawns of millions. Yet Rhys Burton's ancestors had a good eye for water. When the valley's subterranean reservoir goes dry, as it will in a few decades, the last drops will be beneath their land, an underground pocket of water that seeps beneath the farm's soil. The dry pond stores below its surface for a while longer what others so desperately seek.

Rhys Burton's grandmother, Louisa Anna Burton, born in Missouri in 1832, built the first and biggest house on the farm. It must have been sometime between 1879 and 1883. His father added the upper story in 1902, where my children sleep now. The contractor made a mistake on his maiden project, and the ceilings were raised to fourteen, not twelve, feet. Realizing that error, he lost his sanity and stopped building houses. So his unusually tall structure is cool in summer but frigid in winter, and the stairway is, of course, strenuous, better for displaying memorabilia than ascending. And it is hard to buy fire insurance for a clapboard, 120-year-old house, for such homes that are still left are a rare thing in central California, where even ten-year residence is unusual, houses are more often of stucco, and native birth something to bandy about on bumper stickers.

There is neither insulation nor central cooling and heating in the house; all that perhaps might still be my son Axel's task. Cat, my daughter, is the sixth consecutive generation to sleep in the same bedroom, where her grandmother Catherine grew up. Her brother Axel and sister Anna have their great-aunts' rooms. So our kids took up the abode of Rhys Burton's three daughters, who themselves earlier had taken over from Rhys and his siblings, and so on. By 1947, after the big inflation, Rhys Burton had built on the kitchen and added a dining room. My dad put in a bathroom upstairs. Had my mom Catherine lived longer, the two of them, not me, would be living in Rhys Burton's house now.

I've been back here fifteen years, since 1980, and won't leave again. In May of that year, I came home to Santa Cruz from the Stanford University classics department, got in the car, and without notifying anyone, drove to Alma from the coast. The next day I was rebuilding a roof on the fruit shed. So much for nine years overseas and in the universities, so much, too, for my embryonic, but aborted, career. With that three-hour drive I left classics, the university, and contemporary culture for good—or so I thought and would have wanted. My wife reluctantly followed three months later, and we moved into the big house to live with my grandmother Muriel, Rhys Burton's wife, who had been here since she married in 1911. My crippled Aunt Mabel had just died, after sixty-three years of residence in the living room. I inherited the house after my grandmother passed on at ninety-three, and then my mom died but six years later.

My contribution to the evolution of the house of Louisa Anna Burton? A new wing, an extra porch? No. Nothing much at all. Restoration mostly, wiring, plumbing, and painting that makes the house look much nicer than its appearance in photographs from the 1920s and '30s—with one exception. Finally last year I built a stone-block wall all around the house and yard, a 550-foot circuit, 6 feet tall, 3,000 feet of steel re-bar, 50 cubic yards of concrete foundation, more a fortress really than a mere enclosure. No bank, no hoodlum, no broker will make it inside that wall. They may, like the Spartans in Attica, ravage the countryside outside the circuit, battle with family skirmishers on patrol, but the house itself will be safe. Knowledge of the principles of classical fortification and siegecraft (poliorcetics) and ancient Greek masonry is not entirely without use. All the money and time I had went into this wall that will last as long as the house stands and longer still. Its foundations, after all, are two feet thick, reinforced concrete. And so when the land is sold off, built over and forgotten, when the drip system my mother and father bought on credit is ripped out for cul-de-sacs and fire hydrants, the tractors they financed are sold for junk, and I am but a concrete marker in the Alma cemetery that my line does not visit, Rhys Burton's abode will remain for a while longer—behind my wall.

II.

To describe Rhys Burton is to write of an archetype that goes back twenty-five hundred years in the West to the Greeks of the city-state, the venerable but dour rustic who brings his no-nonsense pragmatism and his skepticism to town in time of crisis, to save his republic from the sophists and nuancers who have forgotten where we all once came from. You know him as old Laertes atop his mountain farm in Homer's *Odyssey*, as cranky Hesiod, on the attack against the bribe swallowers in the *Works and Days*, who preaches delayed gratification and the need to be, not the happy, but the useful, citizen. Aristophanes has him the sarcastic hero of not a few comedies, and often he appears out of nowhere as the sanest man in a drama of Euripides. Theophrastus knew the farmer as the boor—this self-employed, cantankerous individual who gives no due to the sophisticate and dapper, who can assume a persona larger than life.

"My God, Rhys is dead, I still can't believe that's possible," so one contemporary of his ancestral cohort told me shortly before his funeral. "It was quite an occasion when he came on into town." Even ornery Tillford, his brother-in-law and self-made prosperous cattleman, said as much. Now wealthy and hale at ninety-five, he still hates Rhys Burton for the years he was poor and forced to accept the Burton charity of job and shelter; yet he too grudgingly confessed on the former's demise in 1976, "Well I suppose he worked harder than anybody but me."

Mr. Burton was bleached white (pink when it was hot), rail thin, with a controlled stutter, terrible temper—and a strange habit of silently stepping forward with his checkbook at funerals, weddings, house raisings, reunions, and the like to pay for what others could not. He paid for the funeral and headstones of Uncle Tillford's siblings and parents; put Tillford's mother, my other great-grandmother, in his own bedroom, now our bedroom, for her last fifteen years.

Ninety-five-year-old, five-foot-one-inch, cash-laden Uncle Tillford, with cigarette dangling out of his mouth, with a glass of VO bourbon in his liver-spotted hand, tottering on his enormous cowboy boots, makes his pilgrimage to my house each week. Every seven days he visits solely to disparage his brother-in-law, two decades after the death of Rhys Burton, "Mister Big-Man, Mr. Smarty-Pants, Goddamn Mr. Democrat, always did things his way, always thought he knew what was best for everybody. Rhys Burton. Rhys Burton. Had to hear from everybody that he moved out to the porch so that my Mother could take his bed. Well, where is Mr. Jesus Christ now?" There's not much left of Tillford now except his boots, cowboy hat, and squeaky voice.

Rhys Burton's vices? One need not get up at five and retire at ten for over thirty thousand consecutive days. Too much watermelon in the summer, a new car every twenty years, a replacement truck every thirty, an occasional tendency to pay a bill twice, to keep a bad orchard too long, to farm a vineyard that did not pay, and to pipeline over and around hills when he should have bulldozed and scraped them down. And I suppose many would say that even his eye for the aesthetics of terrain was not enough, that he fought nature too hard, trapping gophers, poisoning squirrels, digging out all species of flora other than domesticated tree or vine, pruning back wild rootstock suckers in his near century-long struggle to impose his culture on his land. But is not that, after all, what the Romans said agriculture was: man's *cultura* of the *agrum*?

He paid in cash, rarely by check, and handed the money to his seller face-to-face; he never drank, never smoked, ate sparingly, traveled little, married the first woman he courted, and knew no other for the next sixty-seven years. He bought pipeline for his vineyard before carpet in his living room, new trees before curtains, fertilizer in lieu of pants. He irrigated in the morning, cultivated in the afternoon, and read in the evening, everything from the twenty-second book of Livy's history to the saga of the invention of the Fresno Scraper. He farmed with horses, pumped his water by hand and wind, and his agriculture was more akin to agrarians of the ancient Greek city-state than ours is to his. I think he once left Alma and went to New Mexico for four days of a planned two-week trip.

"There's nothing in town unless you have business there, so stay out here and you won't get into any trouble." Aristophanes' Acharnians said about the same thing, and Alciphron's letters of two millennia ago center on a similar theme, the moral superiority of those who toil outside, not amid, the city's walls, who calibrate and temper their sleep, their diet, their dress, and their offspring to the needs of land, not man, who give their all to produce food.

Shame, not guilt, is their benchmark. "Why, I tell you," my grandfather ranted, "if you don't pay a bill, if you get mixed up with beer and cigarettes, or borrow money, the whole town will know of it and they have a good right to wonder just what kind of people we are out here"—this in 1975, when much of Alma was already composed of a *mobile vulgus* hooked on entitlement, and a very tiny elite mortgaged to the hilt to pay the tab for estate-sized homes and imported cars. Yet he cared not at all for the real state of affairs in 1975, whether Alma was 1920 or 1975; whether there was, in fact, no longer any shame in town affected him not at all. When a man has a regimen, an ethic, and a routine synchronized with nature and land enough to feed and protect him, he lives in a world without time, with the smug assurance that all others are as he. No wonder the Greeks called that *autarkeia*, a natural self-sufficiency that was not to be lost at any price—*autarkeia*, the insurance against the fickleness of others and the warp of progress.

Rhys Burton told me his ancestors came from Missouri under mysterious circumstances (a killing, I think, over a Civil War dispute) and camped out here after they bought most of the land from the railroad and homesteaded the rest. Ten years ago, we bulldozed their crude lean-to shack, the oldest structure on the farm. We claimed that it was rat- and bee-infested and in the way when we drove tractor; true, it was all three, but it also had hand-planed wood and square nails and was an attraction for local artists and photographers who said they were in need of ambience.

Everyone prior to Rhys and Muriel Burton has been dead for more than half a century, and, as I said, we know little more than their names and the sepia pictures on the walls. Their saws, hammers, and plows sit out in the barn. Two wagons of theirs are still up on the hill above the pond, the wooden and iron brakes still tied fast. From what I gather, most of their lives were tragedies: cancer, alcoholism, paralysis, insanity, and early dotage, the cost for turning a mosquito-ridden tule pond and some sandy soil into a productive farm, the age-old toll when labor is plentiful and capital scarce. They are buried over in the original Alma graveyard, away from their children, grandchildren, and great-great-grandchildren, away from Rhys, Muriel, Mabel, Louisa, Catherine, my sister, and the others born this century, and few, if any, now know the location of their graves, and none, I have noticed, put flowers there at Easter or Memorial Day.

"Everything but the mosquitoes they packed in here on a wagon"; Rhys Burton also told me that in 1976, the year he died at eighty-six. He meant the stakes, the wire, the concrete pipeline, the pumps, the buildings, the rootings—everything that is not scrub, everything not indigenous to the natural habitat of the San Joaquin Valley in 1878 was brought in by man. Even the garden flora were imported: the fig in the yard, the 120-year-old Missouri walnut tree in the driveway, the ancient varieties of roses, the wisteria that winds under the foundation, all were planted over a century ago to provide some material and psychological relief from the heat and desolation of what early explorers called a vast and horrific wasteland.

And it was a wasteland that these exploiting white people found. Between us and the neighbor there is an odd triangle of land beside the pond and orchard. It's a no man's land of rough terrain where the tractor cannot go, where its small awkward shape precludes planting of capital stock. It sits pristine as it was in 1878 and, I suppose, as it was earlier, in the very beginning, its soil never touched by the plow. It is in some sense an ugly piece of ground, thickets of wild species of willows, an occasional scrub cottonwood tree, jimsonweed, puncture vine, and thistle, its animal inhabitants largely cottontail rabbit, coyote, owl, and hawk, whose clans over the century have learned of the sanctuary and safety offered from chemical and mechanical assault.

But the contrast with the irrigated trees and vines is startling, the difference between an oasis and desolation, between order and chaos: nature's, not man's, land is the eyesore. This anomaly of untouched land is an accidental laboratory, proof positive that the San Joaquin Valley was little more than a desert hellhole unfit for human habitation before the appearance of the agrarians. And so to appreciate the work of Rhys Burton's farm, I walk by this ground each evening, imagine our entire farm as that triangle, the entire county as that triangle, the entire valley as that triangle. It is an ecologist's dream, a farmer's nightmare that would give no life for any other than for the feral, had not small men with horses and strong backs dissipated their youth and vigor in its mastery.

But Rhys Burton was partly incorrect when twenty years ago he spoke about the condition of our land: he, not others, was mostly responsible for its size and appearance, for its radical evolution from the wild habitat of that triangle. He should have explained that, for the first thirty or so years until he took over in 1911, there still was not much and that not much was shared by many. True, this farm was bought, homesteaded, and planted by his parents and grandparents, but its orchards, most of the buildings, the irrigation system, the idea that it was one complete whole, not six twenties to be fought and bickered over by siblings, that entire cargo was his. Still is. He bought them all out, paid off their mortgages, interconnected all their twenties with pipelines, roads, and orchards, made a big rectangle of a family farm that was his and could not be subdivided or sold piecemeal, its parcels not to be untangled by lawyer or farmer.

Blanton, his brother, was brilliant but drank; Orin, his uncle, hard working yet crippled. It is not unfair to say that Anna Lee, his sister, was nice but married wrong; Willem, his father, kind but unbalanced when his wife died young; Schuyler, the other uncle, prosperous but tight. Rhys alone, relatives and friends said—enemies as well—was more than the others' aggregate, and so kept the land when they could not, kept his own alive and fed the others when they could not feed themselves. There were Holts, Zinnies, Tillfords, Dollies, Lorinnas, Elgins, and dozens of other relatives forgotten, a hundred and more, distant and direct, who in the lifetime of this farm sought refuge here, hid themselves in its various dwellings until they were well or strong enough to leave its asylum and venture back out into the fray.

Still, some of you, with reason, might feel sorry for him rather than for those who left. Freed of the land, their successful progeny went on to run businesses, become pilots, fight wars, and make money. Gregarious, cosmopolitan, occasionally even witty and chic, some visited in shiny cars and talked of Los Angeles and New York when Rhys Burton was old, tired, and still chained to his tule pond farm two miles southwest of Alma, California. There is a price to pay for being responsible when perhaps you should not. Eighty-six years lived within less than a square mile has its drawbacks, I concede.

"What a mess," the county zoning officer told me a year ago when he looked up the farm on the parcel map. "With this joint ownership and new zoning laws, it's not worth a damn unless all five of you sell at once; can't be parceled, resurveyed, and the whole thing is in the Agricultural Preservation Act to boot. Whose bright idea was that? Whoever wants his share to get out by himself will never get a dime out of it. Too bad, too; nice location close to town, good land."

But Rhys Burton would not agree. "There's not a place like it in the country. I can irrigate every orchard from any pump on the ranch, even got pipelines under the asphalt road. And my alleyways and hills are like a park." And so they were, because he did not drain, did not level, did not bulldoze down his pond and hills but farmed them as they were, pledging labor and capital as the price of their contoured beauty. I left his name on his mailbox. His diaries line my shelf; his pictures hang on all the walls. He's been dead nearly twenty years, but in all our minds it's still his place, his grand design. We're mere renters and squatters, who hold it and try to pass it on. We won't turn his six twenties into twelve tens, much less three forties; we'll be lucky to keep one twenty-acre parcel—more likely, we'll salvage none.

But the memory of Rhys Burton can cut both ways. When my brothers and cousins fight, when we threaten to sell, when we complain about the poor soil on the south side, the hundred-year-old vines that need to come out, the tangled legal mess of five cousins owning a single parcel, the HUD subdivisions on the horizon, the gang graffiti on the irrigation pipes, the dumped mattresses and TVs in the orchard, the pages of financial statements and forms for the local bank, someone unfairly mentions Rhys Burton. The fighting ceases.

"He made it with thirty people stuck in his house and barn in the depression when raisins were thirty dollars a ton." End of argument, *ipse dixit*; therefore, we can easily make it when raisins lose 60 percent of their value in a single year?

"He was broke at forty, with a crippled daughter, a wife with a ruptured appendix, a mortgage, no capital, and died at eighty-six with money in the bank and his land paid for, his daughters graduated from college." End of argument; ergo, being forty-one in 1995 with his land and no money is not so bad after all?

"He had a heart condition at twenty-one and could not even buy life insurance." End of argument. *Quod erat demonstrandum.* Therefore, our bad backs, loose teeth, and stone-clogged kidneys bring no sympathy as the price of our struggle?

His lesson to his progeny, besides the tired triad of "don't move, don't divorce, and don't change jobs"? Simply put, it was not much more than four precepts: (1) Go away to college outside the valley and learn a profession, preferably law, medicine, or teaching. (2) Come back and live on the ranch among your own kind. (3) Never sell. (4) Don't let more than one person farm at once. "The ranch won't support you all," he thundered ad nauseam.

But is that not against the agrarian ethos? Should not *all* yeomen farm their ground, do their own work, raise their children at their side, the more in the patchwork the better? Perhaps, perhaps not. But my grandfather was a realist, not an idealist. He was not a mere romantic, Horace's *laudator temporis acti*. He himself was on his way to college in 1911 when his mother got cancer and his father lost his mind, and so for that catastrophe alone he stayed. But he knew from farming on the edge for seventy years that the family farmer in America was eventually doomed, that the tradesman of the 1920s grew to the broker of the '30s, expanding to the cold-storage owner, shipper, and packer of the '50s and '60s, bloating to the corporation of the '70s until reaching fruition as the financier and insurance company of the '80s and '90s, a logical, unwavering cycle that would know no end until the farmer was no more. "Get big or get out" was their cry.

Still, the farmer and the land—could they not be two different things? If one sibling were to farm, but two or three were to earn a living off the farm, could not all live nearby together? And so my mother did not farm, and my father did not farm. They came back from the university, and they lived next to Rhys Burton. That in and of itself was a wise and noble thing, but already by the 1940s their decision to seek employment elsewhere was clear proof that agrarianism was dead in America.

Our farm explains everything that was both good and bad in my parents' life, everything both heroic and tragic, easy and difficult, intelligent and dumb. They rose at dawn, drove to Fresno and worked, came back in the evening, and listened to Rhys Burton and my grandmother Muriel. They did his income tax, his welding, his workman's compensation, and his tractor repair. They canceled dinner invitations at harvest, listened to the weather reports at work, and worried about sugar contents, stormy weather, and Ford tractors when those of like kind talked of affairs, divorces, stepchildren, private schools, Europe, and golf. They bemoaned the road dust and chemical spray but walked through the vineyard at sunset. They worried that we knew little of the world outside, but told us that all that early digging and shoveling would save us when others failed. They were embarrassed about their tiny house but beamed when they showed guests the surrounding 120 acres of lush orchard and vineyard. What was a promotion, a salary hike, or chance for transfer when frost had nipped vine shoots and rain rotted raisins on the ground?

If Rhys Burton was a nineteenth-century man, with all the limitations that rubric implies, his daughter Catherine, my mother, was twentieth century and more. If he was to save his farm through sheer hard work and agricultural acumen, she was to do the same through the application of off-farm capital and mastery of the statutes of the world outside the tule pond. She was the first woman student-body president in the history of Alma High School, the first female superior court judge in the annals of Fresno County, and the first female state appellate court justice in the Central Valley of California, taking to heart her father's dictum that though agrarianism per se was dead, at least it might be a rock for an existence elsewhere, as working in town was in turn a salvation for the farm.

Feminist historians, I imagine, argue that women have kept the American farm alive and have not received their due; behind every man with a hoe, they say, was a woman nursing, washing, and counseling. With great reluctance and some trepidation, I confess that this more sober, more balanced, less epic picture of agrarianism is true. When Rhys Burton's vast latticework of subterranean irrigation pipe began cracking—the depression-era contractor had skimped on the quality of cement—Muriel rescued him from his own greater depression: "Hold on, this too will pass." When sons and nephew were beaten down and smashed by $450 raisins and $3 plums, Catherine began weaving a complicated tapestry of mortgage, personal loans, life insurance, and annuity to resuscitate her men. When we fight and debate now how to live on land that takes rather than gives, our cousin Inger, the fifth and silent co-owner of Rhys Burton's farm, drives down to adjudicate, compromise, and arbitrate. And when I talk of Rhys Burton or of his father Willem, I would be well advised to remember that it was Louisa Anna Burton who bought the land in the beginning.

But unfortunately for Catherine, perhaps luckily for us, she did not put the lucre that accrued from the law, that reputation at the county bar, to the comfortable life, to an estate in North Fresno, a vacation home in the mountains, to island vacations in the Caribbean, and to the accepted pelf and leisure that accrues to the honored barrister from consultations, trusteeships, conservatorships, and executorships. No, all her time and meager funds went to Rhys Burton's farm. Thus, the visible, material remains of her life lie locked up irrevocably and nearly invisibly with the success or failure of this farm and those who work it. Some might say it will end when this farm ceases as well. She lived in the house that my dad moved onto the place in the late 1940s, my twin brother's house now. Again, what money she made she put in the farm.

My dad did the same. His father's place of forty acres is a mere ten miles away in Lundburg, well over a hundred years in the family, too, and so he is the third in a line of farmers. My other great-grandfather, Knut Hendrickson, came from Sweden in the 1880s and homesteaded some of the best land in the entire valley, just ten miles east of his contemporary, Willem Burton, who at the time was busy scraping inferior ground around his tule pond. Axel Hendrickson, my father, was willing to farm Rhys Burton's place and his own nearby but didn't. In fact, he ended up farming neither. That's a long story and involves hurt feelings and is better left alone. Suffice it that Rhys Burton's other son-in-law, my cousins' dad, my uncle-in-law, wanted it all, got it all, and soon wanted none of it at all.

Instead, my dad went away to college, came back, and did not farm. He ended up with a good job as a junior college administrator, a job he did not particularly enjoy, one that farming and military service—which, as the Greeks tell us, inculcate heroism and sacrifice—had made him entirely unsuitable for. So he became the son my grandfather never had: mechanic, welder, bulwark when neighbors threatened, stole, and cajoled, master repairman, and tractor driver; a very good person to have around when the district ditch tender periodically decided to short Rhys Burton of his last day on his seven-day turn on the irrigation canal. Again, my parents put all their education, their salary, their capital, their vacations, their profits, their lives into a farm they did not farm, into land they did not own. "A bad investment strategy," their associates in Fresno advised.

True, my mom finally owned some of it, but only for her last thirteen years; my dad, never. Catherine did not go on junkets with other judges because we needed a tractor that could pull a scraper. We bought a nice one, too, Big Boy we called it, a magnificent used Allis Chalmers. Dad never bought a new car because we needed a drip system—thirty acres of it; we helped put it in ourselves. And in the complex world of finance and capital, the murky world of interest, deferred income, and dividend that my mother mastered, her subsidy goes on from the grave. Catherine finally put half her income in investments for retirement but died of a brain tumor two years before she was almost through and ready to relax on her farm. Her money, a modest income, and the life insurance went to my dad, which means it, along with his own retirement, went to the farm. He lives down the road now, like a pauper, in a one-bedroom farmhouse. And his money goes to the bank, nearly all of it, for the farm he never farmed and the land he never owned. He told me yesterday he thinks he can sell his pickup and pay off the rest of last year's still unpaid debt.

My two brothers, my cousin, and I got it almost right—at least three out of four of Mr. Rhys Burton's canons. We left for college. We came back. We did not sell. Our mistake? Unfortunately, the most crucial of all the commandments was forgotten: we all farmed—and that is a foolhardy thing to do now in America if you wish to save the land. Logan lives down the road on the east end of the farm, next to my cousin, Rhys. My dad is behind them in a house next to and smaller than Jacinta Guevara's. I'm a quarter mile west in the old main house. My other brother, Knut, moved to the coast.

Still, as you will see, we almost pulled it off, we three brothers and our cousin did. We farmed together, built, planted, and prospered, paid our debts, bought equipment and made a tidy profit until not long after 1982, our last and best year. Rhys Burton was right, after all: go to college, live on the land, don't sell, but also don't all farm. I went to teach, Knut went to teach, and Logan, I'll wager, is about to go to teach.

When we argue about the money and the losses, about how the entire farm became unraveled, my brother has a tendency to say, "Now wait a minute here, mister. Had we got just a dollar or so more for a box of fruit, we wouldn't be talking like this at all. We're fighting over nothing more than a silly dollar more a box." He's right. But I also tend to say that we are owed far more than that dollar on every box; five dollars of profit and more still we are owed per lug for what we have gone through to grow such fruit. But in this world we will not see that dollar, won't see even the twenty-five cents more that would make all the difference between this farm's life and death.

As my parents learned, it is not easy, these parallel lives—at work in one century, at home in another. But, you interject, does not a farm inculcate hard work and prepare one for the softer world in town? It does. And so in college we all studied every evening, received our degrees, got good grades, as if we were perpetually rolling trays before a big rain. Surely a farm teaches you to be versatile, to weld in the morning, to irrigate that night, to prune tomorrow, to view the world pragmatically as challenges to be met, countered, and overcome, never to be averted and sidetracked, prerequisite values that any employer might desire? It does. We all took our biology, read our literature, avoided the campus Marxist and right-wing nut alike in pursuit of degrees that made us employable. Finally, does not a farm forge family ties, the notion that blood above all is to be honored and protected, feuds, divorces, and alienation to be avoided? It does. We all went to the same university, the University of California at Santa Cruz, the most affordable college and the one closest to the Alma farm, lived in the same house, had the same friends, drank the same beer, dated the same girls, stayed married to the same women of the same locale.

But the farm is still more a disadvantage than an advantage even to those who live there but do not always farm. It is no longer of this world, this late twentieth-century American world, and so by necessity more often farm life ensures havoc, not tranquility, when its denizens venture to the outside. Farming is heroic, a midnight dash to get water on blooming fruit trees when the spring nights plunge below freezing. Farming as a natural enterprise is tough, the need to disk and harrow with the flu, prune with the sprained ankle, pick overripe fruit with the stones in the kidney. Farming involves deed, not mere word, reality over theory, substance over brag and big talk, for ugly trees can produce beautiful fruit, rain comes when it should not, and kindly neighbors will press you until stopped. And so of all the trades and professions in America, farming turns out to be the worst preparation for the clerk in the department, the professor in the classroom, the judge on the bench: It insists that you speak up when you should be quiet, jump in when you should step back, hack when you should nick, and say "no more" when you should nod yes.

III.

"All men," Emerson wrote in an essay on farming, "keep the farm in reserve as an asylum where, in case of mischance, to hide their poverty—or a solitude, if they do not succeed in society." My own memory of 1984—when I ventured off the asylum as a part-time lecturer of Greek and Latin at the nearby California State University, Fresno, campus—was, even given my lowly academic status, the sudden change in value of my work in the space of literally a few days in early September.

My income at first was about the same ($8,300 per annum; but there were to be pay increases in this world outside of farming, off-farm monies to pledge to the bank against on-farm losses). My title changed from "one of the Hendrickson boys out on Rhys Burton's old place" to the pretentious "Dr. Hendrickson." My classics B.A. and Ph.D. degrees were now something other than the butt of repeated jokes from fellow vineyard pruners about a near-decade of schooling, my twenties in their eyes completely lost and wasted. Even the scruffy visage of disheveled hair, clothes, hat, and sinking trousers no longer caused polite embarrassment to the urban onlooker (did not Euripides say, "not much to look at—a farmer") but rather indicated metaphysical otherworldliness—all as reward for a workweek shortened by half and a monthly paycheck whatever the weather, the market, or the needs of others; a life now characterized by a physical environment cool in summer, comfortably warm in winter, free of dirt, grease, toxic chemicals, the random violence of aging machinery, and the standoff with the occasional disgruntled farm employee.

New associates suspicious of a farmer in the classroom hoped that I could in time turn out to be a supportive—and normal—colleague. But to families and friends back in Alma on Rhys Burton's farm, even though I was not drawing off its vanishing sustenance, I had shamelessly opted for little more than a prolonged vacation. I apologize for it still, over a decade later.

The most difficult drawback on campus, as I then understood it, was how to avoid ingrained habits, offensive and uncouth all: field urination; commonplace profanity; the use of the sleeve in place of the napkin and handkerchief; blunt appraisal of the coffee break, weak administrators, the general claustrophobia of and dislike for forced hours in the city, and the inability to order the unpleasant off one's property. Skills of the last years, sharpened by experience and the cause of some pride in their reacquisition, were now utterly valueless. Tractor maintenance and repair, pruning, grafting, irrigating, rudimentary carpentry and plumbing were a joke. Knowledge of weed species and insect pests, experience with proper raisin moistures and microbes, the principles of on-farm food storage, a mastery of fertilization and chemical use, the ability to grow food in volume—in the unreal world of the urban workplace, all were superfluous. These experiences might have been laughed at, if not despised, if they had been known. On the few occasions when I related the cosmology of agriculture to my newfound university colleagues, men and women who had ostensibly devoted themselves to unblinkered inquiry, to an interest in understanding the world about them, responded with a polite lack of interest.

Is it not odd to rise at dawn with Japanese-, Mexican-, Pakistani-, Armenian-, and Portuguese-American farmers and then be lectured at noonday forty miles away on campus about cultural sensitivity and the need for "diversity" by the affluent white denizens of an exclusive, tree-studded suburb? Associate with, work with, live after hours, summers, and weekends with elderly men and women—white and brown, family members and neighbors alike—who work routinely sixty- to eighty-hour weeks, often (as the year's final proceeds in retrospect tragically demonstrate) for less than the minimum wage; then Monday through Friday hear lamentations by recent Ph.D.'s over the injustice of our department's "burdensome" tenure requirement of but a single published article.

Or, in the recent California drought, to hear at urban gatherings boasts of households that have ingeniously invented ways of consuming (properly so) only fifty gallons of water a day and that night to drive back home to coax seven thousand gallons *a minute* out of aging pumps and failing wells to save six generations of trees and vines. To see an entire harvest of raisins—two hundred tons of them drying on 100,000 paper trays beside the vines—"float" down the vineyard row as miniature rafts, all ruined in an unseasonable September storm, and that same week to be consoled by colleagues over the denial of a small research grant: that contrast is nonsensical.

On it goes daily in this last age of American family farming, lives at the poles of the earth, the divide between the farm and the city numbing if one unwisely stops to ponder all of its multitudinous and often bizarre manifestations: the spectacle of hard-working Mexican-American farmers eagerly adapting, modifying, but ultimately accepting contemporary culture, as they name their children Matt, Jeremiah, and Nicole, plastering American flags over farm machinery in the wake of the Gulf War; while at the same time, too many of their self-acclaimed representatives—the affluent bourgeois potentates in the university—*re*christen themselves Pedro and Tomás for the scraps of the contemporary academic feast. Farming, and the people who farm, can (devastatingly so, almost cruelly so) set off, frame, make sense of the nonsense present in the modern American workplace.

Most then do not wish to leave their fields of dreams, to venture back into American society—not my grandparents, parents, siblings, or cousins; not me either. They pour their hoarded but vanishing capital into a quagmire that has no end, these few thousand left still in California with their own creaking century-old houses, south sides, tule ponds, and brains stuffed full of Rhys Burton–type wisdom. They know that family farming is over in America, and they do not give up.

They do all that and more because they have seen the brave new world of nonfarming urban America, and of it they want no part. Increasingly I have wondered why that is so.

CHAPTER TWO

THE RAISIN COSMOLOGY

Our farm was and still is dependent on raisins. Figs, peaches, apricots, and about anything else will grow in the valley and grow well. But many of these species will also thrive elsewhere. Raisins alone, the ancestors learned, can make it here. Oh, when they go bad, we diversify—in some decades shrinking down to forty acres or so of vineyard, in others slowly expanding back up to nearly a hundred. But for good or evil raisins have been our salvation—and then again for the past twelve years our death knell.

There are no raisin plants. Sound silly? A Greek professor (he could not have thought much of farmers in general or me in particular) once used that precise phrase "raisin plants." In 1976 he said that he had heard from the other graduate students in the classics department that my family had lost 180 tons of raisins during a bizarre tropical storm. He asked—smirked really—how many tons my raisin plants grew, as if I were some type of crazy raisin plant farmer.

Even if there existed species of raisin plants, I have since thought, wouldn't farming them make a whole lot more sense than what he did: reading one play of Sophocles (the dramatist's worst) over and over for twenty years, having five to eight unpaid graduate students in each year's identical seminar recheck his never-to-be-published theories about whether the play's chorus had actually split into two—or was it four—subgroups? Wasn't an $80,000 salary for such toil a lot sillier than $7,000 a year for farming "raisin plants," for providing a half million people with their yearly raisin intake? True, classics and family farming have both now become dying, unsustainable professions. But at least there is an element of tragedy in the doomed industriousness of the latter. There is really none in the former's elite and calculated sloth.

But raisins are dried grapes, not plants. They are most always now dehydrated Thompson seedless grapes, the green, oblong berries you see fresh in the stores, in their finished form now the obnoxious claymation commercials on American television. Muscat, currant, sultana, and fiesta grapes have all but disappeared; my great-grandfather Willem's sultanas and muscats were grafted over half a century ago to the ubiquitous Thompson seedless, their history known only by the occasional sucker that some years sprouts out of a gnarled Thompson trunk.

Processors and consumers alike say they no longer like those tiny, pitted, or black raisins but instead prefer them uniformly plump, purple, and seedless. Americans want their food without the complication of tough skins or seeds. How California's four thousand Thompson grape growers make their raisins is peculiar, a procedure at odds with all logic and sense, a rejection of the whole notion of technological progress and constant American agricultural evolution, a dangerous gambit completely antithetical to the sober, conservative farmers who are necessarily its perennial queer players.

California viticulturists produce raisins the way they always have since the vines were first introduced to the San Joaquin Valley in the late nineteenth century. They simply pick the bunches and spread them on paper trays beside the vine to dry in the sun for about two to three weeks. We harvest and dry no differently than Rhys Burton, who did it identically as his father and grandmother did. There are no mechanical pickers, no enormous gas dryers, no retractable tents to cover the fruit when weather threatens, no concrete drying yards, sprayed-on chemicals, or miracle dehydrating dip. The American propensity for machines and behemoth contraptions has not worked in the raisin industry, at least not yet. The equipment prerequisite for raisin production has always been and remains a ninety-nine-cent grape knife, some paper trays, strong arms, a back of iron—and cloudless weather.

Within that simplicity, as I hinted, lies also an insanity, a madness inherent in the whole raisin community. What else but derangement is it when a family places its entire livelihood, the bounty accumulated from generations of work, in the dirt, either to be wiped out in an afternoon rain or quadrupled in a day, depending on whether the metamorphosing grapes are exposed or stored safe in the barn as finished raisins when unforeseen rain strikes? Nowhere else in farming, much less in any other profession, can the outcome of an entire year *after harvest* be reduced to the caprice of a few hours. Then unforeseen lucre or ruin hinges largely on how many of one's kindred growers pick too late and see their crop putrefy, diminishing the always present oversupply. In rain years, atypical though they may be, it is too often a zero-sum game of the traditional peasant. Your profit often depends on someone else's perdition. "It's not enough that I get by," one neighbor muttered to me, "You guys, you have to fail."

The ideal scenario, I suppose, is to pick early and have an early rain: your farm's raisins are dry in the barn, everyone else's are rotten. In theory, their catastrophic privation drives up the value of your scarce crop. Brokers pay for what they cannot easily have. On the other hand, the worst nightmare is to harvest last in late September and get hit with a storm. Everybody but foolish you saves his crop; raisin prices still stay flat, but you alone get nothing. "I bet you're happy I lost my crop," barked an angry wealthy fool who picked late, waiting until the last drop of sugar was in his grapes. "Not happy, but not unhappy either," was about all I could offer.

Don't think that we early, rain-scared, capital-exhausted pickers are simply gamblers who desire the ruin of others. Many of that late-picking bunch are worse. The majority of September's harvesters are the wealthier farmers, the more established (often the board members of the Valley Girl raisin cooperative), who can take a loss but demand a constant profit. These dour-faced always want the top quality, sweetest, and heaviest raisin for themselves *alone*. They play a different (and better) set of odds that it won't rain, hedging that the poorer early pickers of August, for whom a single year's rain means instantaneous liquidation, are driven by poverty to give up their grapes' weight, panicked to let its immaturity evaporate as water into the summer atmosphere, reducing the overall size of the valley's crop, emphasizing the superiority of their own later, sweeter September harvests. Each day the crop of the late men stays on the vine, they sit confidently in their pickups as their grapes gain weight. No, I prefer the early picking cohort, not the smug visages that appear in *Vine Grower* magazine or *California Farmer*, whose inherited capital allows them to misinform us how they still manage to profit in an all-but-gone raisin industry.

There is also, I know, an undeniable beauty, a simplicity to the entire system of raisin making, quite aside from the danger and morbid interest in the fall of others, that explains its existence unchanged for more than a century. Drying raisins, as I said, is an aboriginal activity. It is one link, about the only one left for the modern California cultivator, with a cadre of farmers now gone, the mustached faces in faded brown photographs on the staircase wall who first harnessed the waters of the Sierra, planted the vines, and learned how to dry raisins—a process of leather, wood, iron, and flesh born in the world of horse and wagon, where sheer muscle, logic, and ingenuity overcame the absence of technology, a profession handed down nearly pristine from father to son over six generations. Rhys told Catherine about the baffling paradox of when to pick grapes; she lectured us, but he learned the ideal from his father. And Willem? I suppose he and his mother taught themselves, for it was altogether a strange thing to plant vines amid scrub, to produce raisins in 1880 for those few who knew what they were.

Each September agrarians, be they twenty or eighty, experience the same fear (and thrill) as all put all their capital, their land, their houses, themselves on top of the dirt. There is no retirement, no well-earned security and complacency in this raisin business. The vines sprout every spring. Their crop matures in late August. The grapes must be on the ground in September, each year, every year, all years. "I've about had it," rail-thin, stooped Rhys Burton sighed about a year before he died, "and I tell you I could lose everything I earned these eighty-six years in a single afternoon. I'm glad there's not many left." You never win. As long as the vines remain in the ground, you never cash out and call it quits. You simply try to stay on the back of your raisin beast as you pass its reins to those you hope you have trained well enough to ride it for another forty years.

Californians today can grow and harvest raisin grapes—save for the substitution of combustion for horse power—just about the way it was done 120 years ago. No wonder that local agricultural economists lecture on its backwardness (as if the cost of production, not the price received, had ruined us). The university can never let a good thing be. They have no patience with labor-intensive activity, the vulnerability to the elements, the absence of hydraulics, belts, pulleys, and the blinkered unconcern with lasers and computers. Out of disparagement the university tinkerers constantly fabricate quirky-looking contraptions, sleek, brightly painted (in John Deere green or Ford blue) chariots that mechanically pick an experimental vine row or two at great cost, but thus far they have inevitably failed in the grimy anarchy of the real vineyard. Only these machines' catchy names impress, these Raisin Wranglers and Grape Getters. It is hard, remember, to match the craft and proficiency of the ancestrals' bare-bones protocols of turning grapes into raisins, of mastering heat, dirt, and moisture to produce a natural and inexpensive candy. Their economy of purpose, their routine from picking to boxing, was driven on by the instinctual need for food and shelter, not the maximization of mere profit. And so these tiny men were often smarter, better than we. We still follow their Spartan customs and put our grapes right on trays in the dirt.

Drying raisins can also be visually and aromatically an arresting experience: thousands of tiny mauve, shimmering trays over the white sands between lush green hedges, the whole countryside smelling like an enormous bakery (which, of course, it is). Visitors are astounded when they walk out into a vineyard and sniff grapes dehydrating in the air, when they sense that one hundred acres of grapes have been converted into four hundred bins of raisins in the space of three weeks, half a million pounds of raisins cooking outside the kitchen window.

Once the berries reach the ground to fry, the frenzied farm activity of an entire year suddenly skids to a halt, collapsed and silent. What can a raisin farmer any longer do in his field when his year's work is scattered there all over his ground, a thousand trays to the acre? Irrigate over them? Cultivate through them? Fertilize or spray on top of them?

A strange calm descends, as entry to the vineyard is barricaded by the year's investment beside the vines. Nature mandates a pause as the grapes slowly cook into raisins. Meddlesome toilers of months past, their tractors, sprayers, and harrows parked, are now for two weeks relegated to nothing more than pitiful and helpless bystanders, who mill around their shops and barns looking at the sky, listening to their radios, waiting for the day's satellite pictures on the evening news—as if they could really do much with their just-picked grapes if it *did* rain.

At a precise time the drying grapes are rolled to cure in their paper trays between the rows. Roll too early, they liquefy into mush; roll too late, and they caramelize into rocks. In some years you roll on day ten, in others on day twenty, calculating always the degree of sugar in the drying grapes, the humidity and temperature of the air, the texture of the soil, the extent of the vineyard's canopy, the size and nature of the bunches, the type of paper in the tray, the size of the crop, and the days needed to remove the rolled trays from the field. The wizened farmer then visualizes the perfect scenario, the ideal form of how it all should go: from exposed, to rolled, to boxed raisin.

But that is only the naif's dream on the other side. A vast canyon of uncertainty lies in between—the reality of men who do not show up, of tractors that stall, of weather that turns fickle, of weakness before panic and second-guessing. So when the farmer tallies all these factors and decides to roll his trays, he needs the men immediately—a half day too late and the harmony is lost, his profit evaporating into the air.

Your ally, the sun? It can ruin a year as effectively as a tropical monsoon. In September 1977, 105,000 trays were ready at six A.M. By nine A.M. six instead of sixty men had arrived for work. By noon it was 105 degrees, and the skeleton crew had retreated home. By five P.M. a record 112 had scorched everything in the field, cousins, brothers, parents, friends on their knees in vain scooting rolls of caramelized rocks under the shade of the vines. Fifty tons—365 days of work—went up in a day's sizzle. "We did better last year when the rain took nearly all the crop," my cousin figured.

No wonder that the unsure waiver and sometimes collapse. Twenty-five-year-old Keith Frontierre down the road once in 1982 rolled too green, unrolled, rolled again too green, unrolled, then boxed too green, then spread his crop to dry on black tarps in his barnyard, only to end up with caramelized pebbles when an unforeseen heat wave arrived over the weekend.

And solitude is as lethal as youthful folly. Seventy-five-year-old Vaughn Kaldarian, thirty-second-degree raisin master, veteran of fifty harvests, had too many specters in his brain—sudden rain, unforeseen heat, rocky raisins, mushy grapes—and too few confidants in 1982. Determined to maintain his steely facade, he saw no good reason to cave, to roll green with a few clouds against the Sierras. And so all day long he drove by grim, unmoved by the mounting clouds as we, his foolish neighbors, did roll for hours on end. Kaldarian determined not to put fat purple grapes inside paper; he had no belly to pay for rolling on Friday what was to be unrolled on Saturday. But at two A.M. the water came. At six A.M. a shaken Kaldarian, slumped over in his pickup in the alleyway beside his sixty acres of raisin archipelago, offers a meek, "Should have rolled, but I had no one to talk with."

The dust created by terracing the earth, kicked up again by the spreading of millions of trays, stays in the stale air of the San Joaquin

Valley for days, a fitting denouement for a stagnant ozone summer. If a farmer can stop to notice, the ensuing polluted sunsets in the late evenings are almost pretentious. The pastel skies are certainly unmatched by anything on the pristine Pacific to the west over the mountains. As hybrid spectacles, not wholly natural, not entirely fabricated, they are the proper residue of human and natural energy. But often it takes the rural dispossessed, the crippled, and the senile of the valley, whose contented brains are not cluttered with the ceaseless drive for survival and success, to see what others, the daily dust makers, miss before their eyes. When all else had gone from the mind of my ninety-three-year-old grandmother Muriel, when she could hobble only out to the front of the house to watch for the ghost of the "evening stagecoach," she still in her dotage directed us, us the oblivious, to the September sunsets on our front porch of a century: "Call your mother on the wire to look west, I'll go get Rhys [long dead], the sky's all on fire out yonder."

But the powers of the harvest sun are utilitarian, not merely aesthetic. They dissolve 80 percent of the grape and change its color, leaving only the dark skin and a sugary core inside. A hundred-acre vineyard averages about ten tons of fresh grapes per acre, about a thousand tons in all. Picking a farm's two million pounds of fresh Thompsons for the winery or fresh consumption is work. It is a monumental feat of logistics, a breakneck endeavor of picking, grading, and trucking. Yet for a raisin farmer those same grapes can also quietly, effortlessly dry down, disappear to one-fifth of their original weight, from a thousand tons fresh to two hundred dry, without all the equipment. If their dry price is five times higher per ton than green, the raisin farmer can make the same amount of money as the fresh grower—engaging the sun, not man, to carry off 80 percent of his crop's substance. Yes, the farmer may seem to be idle for three weeks, hovering around his picked and drying vineyard. But millions of free heliac BTUs are dehydrating his crop before his eyes. Is this not the greatest application of environmentally correct power in America today? Is this solar drying not either completely unknown or disdained by the urban zealots who advocate just such a mass application of natural energy wherever possible?

Bunches are picked off the vine and then laid right in the vineyard rows themselves, once the ground has been tractor-terraced into slopes that face south. The burden of the entire harvest is moved by the muscles of the pickers over a distance not more than two or three feet from vines to the papers on the ground. The sun, as I said, does the rest. Grapes are dumped and spread ("laid") on two-by-three-foot papers ("trays") and then left alone for anywhere between twelve and twenty-one days, depending on the weather, the humidity, and the harvest season (anytime from August 20 to October 1). I have seen our grapes dry into raisins in eight days and I have seen them never dry at all. They have been safe in the barn ten days after picking and they have been disked up as fertilizer in December; they have fed families of ten and they have ruined an entire clan.

The intense heat between the breezeless vine rows (100 to 120 degrees and more) cooks most fungal spores, fries worms, and desiccates a very odd assortment of predatory insects that crawl out to seek their (last) sustenance on the trayed green bunches. The most effective pesticide and fungicide in the world, as the early raisin farmers knew, is the sun's microwave. Far better it is for killing bugs than the old malathion-soaked tray or the airplane that gasses the trays in your field as it douses your family at dinner. The nearby Mohinder Kapoor used to aerial-spray his 120 acres of drying raisins with Captan dust, in the process covering our entire house with a white powdery film. It was a purported carcinogen, yet once he tipped me off, I noticed that the fungicide kept the mold off the walls and curtains pretty well.

After the grapes are *about* dry (no longer oozing juice but before they burn into "rocks"), the trays are rolled into balls ("biscuits"), left to cure in the field for two to five days ("sweating"), and then thrown into vineyard wagons and hauled to the barn. Dumped and screened into boxes or bins ("boxing"), they are then trucked to a commercial packer, who again shakes out debris (sticks, rodent excrement, leaves, rocks, rot, stunned mice, reptilia, etc.), stems the raisins, and then washes and pours them into cardboard boxes for the grocery store or food processor. Raisin packers are dubbed "processors." But the crop long before has been processed from grapes into raisins by the farmer. For all their bluster and exorbitant charges, they merely stem and wash the already finished product. They are really mere brokers who buy from the farmer and sell to the store or food company at quite a profit. Bus Barzagus our neighbor and his son built a raisin stemmer and washer. "Guess what," they chuckled on its completion, "we are 'processors' now."

So language reveals a lot now about California agriculture. The change in agrarian nomenclature and usage is not explicable by the age-old antithesis between the urban refined and the rustic vernacular, between old Aristophanes' dapper *asteios* and reeking *agroikos*. No, it is the long-awaited arrival on the farm of America's odd marriage of social science and finance. The former seeks to reassure the latter that it is something that it is not, something corporeal and material, not vaporous, something humane and civilized, not covetous and rapacious. Mushmouths in white shirts now transfer their liturgy of spreadsheets and Lotus to the growing of food.

Poison is no longer even "pesticide" but can become a more reassuring "material." Spraying and dusting are cloaked and sanitized by "application." Even "liquidation" is not a kind enough word for foreclosure; it has passed over to "consolidation" and "reorganization"—a "process" not to be remedied by going over the books but a more esoteric art of reading your "portfolio." A job in town is broached under the rubric "alternative occupation," as the loan faces "termination" due to "equity erosion" and an absence of "off-farm capital." The farmer himself is called the "grower" only by his lowly workers—better to be dubbed "management," his work "production," by the more financially correct.

His trouble with the bank, poor prices from his broker? Mere "capital scarcity" and "revenue degradation," with "substantial exposure" to any relatives who foolishly signed on the dotted line. "Viability" or lack of same is the bank's catchword, as legions go broke due to a "surplus of product."

In times of such crisis, the Greek historian Thucydides said that "words had to change their ordinary meanings and to take those which were now given them." The Athenians too played at this game, for Plutarch tells us they "called whores 'companions,' taxes 'contributions,' the garrison of a city its 'guard,' and the prison a 'chamber.'" Herein lies a clue to the purpose of all this linguistic metamorphosis. To frame agriculture in terms applicable and welcome to its new guardians in law, finance, and insurance, poisons, bankruptcy, and ruined lives must not conjure up odious images. Better to envision farming and the historic grime of farming as no more discordant than the arts of retirement planning, prenuptial agreements, and stock divestiture.

And so of all the bequests that agrarianism has given America, in its eleventh hour it has offered up its penultimate and most precious treasure, the preservation of its language. Alone now, farmers safeguard the American idiom, concrete, real words that once shocked and then infected the world. There are still bosses and punks, cowards, cheats, and no-goods on the tongue of the yeoman, still snickers and guffaws roll out of his mouth when the television screen delivers a blathering fool in management, government, or law, fat-headed and preened with "the impact is considerable," "at this point in time," "to the best of my recollection," "the successful applicant must," or "the meeting occurred but was not substantive."

The California raisin industry endured or (at least until the last decade) succeeded, I think, because of four simple prerequisites: climate, water, labor, and agricultural technique. If just a single component is wanting, if just one misfires, the harmony is lost, the entire enterprise fails, and the year is gone: 1976, 1978, and 1982 were forfeited to rain; 1987 to 1991 were drought years, when the pump bill and the dry well sucked up any cash unclaimed by the bank; labor was short and hard to get between 1976 and 1980; after 1983 the price was so bad it no longer really mattered.

My grandfather's diaries go back fifty years and tell a similar story. Even the threat of light rain on drying raisins mobilizes the entire clan. Two entries on Monday and Tuesday, September 16 and 17, 1957, read as follows:

> Cloudy, light wind. Had light sprinkles of rain. Buford took all our help on the place and rolled everything that had been turned. The 14 acres on the West, 20 acres north of the road and the vineyard east of old garage had not been turned and were not rolled. Manuel irrigated Buford's peach trees and the East block of late Santa Rosas. Our two grandsons Rhys Andrew Jensen and Knut Hans Hendrickson began school today in the first grade.

> Clear, but cool. Buford will take all the help and now unroll cigarette rolls of raisins. Manuel will continue to irrigate Buford's peach trees. Joe and Chester will unroll cigarette rolls and roll biscuit rolls of the dry raisins on the south side of the place. Delmas will turn wet trays east of old garage.

So rain is not just an outside variable. It is instead a specter that lies inside a farmer's brain, gnawing at him constantly. At summer's end, does the raisin farmer hold off picking his grapes until his crop reaches full maturity (20–22 sugar points, called Brix), ensuring a top quality, heavy raisin (raisins are sold by weight)? If so, this mandates in most years patience, a long wait for harvest until after September 5 or even later—and a subsequent drying period of at least twenty-one days.

For the farmer who seeks the best quality of raisins and thus the best price, the cost of delaying his grape harvest for a week or two (until mid or late September) is occasionally high and one he is bound to pay over. For each ensuing day that the grapes are left to ripen on the vine and thus gain sugar and weight, the chance of fall rains increases commensurably. For each of those days the drying time on the ground lessens, as the daylight hours wane, the darkness lengthens, and the humidity rises. Each day that the grapes stay on the vine unpicked gaining crucial sugar, the chances *decrease* that his fruit will ever dry into finished raisins.

"Have 'em all in the sweat box by the equinox," my grandfather used to say of the need to get the raisins dry, out of the field, and boxed by around September 21, "and you'll usually make out all right." Compare the similar advice on harvesting from the modern dry technical manual: "Harvesting should start when the grapes reach 22° or 23° Brix or September 1, whichever comes first. If the grapes have not reached Brix by September 1, it is an indication that the vines are carrying an over-crop and measures should be taken to correct the condition in future years."

Sometimes the scholarly advice that "measures should be taken to correct the condition in future years" is of no solace at all for the poor farmer on the brink of disaster on September 1 with an enormous and immature crop, with berries still unripe from a cool summer, and with too many clouds now in the sky. For him it is not "future years" of the university raisin manual but the here and now of overdue bills and a mortgage—and thus now no choice but a throw of the dice. For this gambler who wishes—must wish—top quality *and* maximum returns (the ideal but impossible combination) on his year's work in lieu of absolute safety, who must therefore wait *too* long, the ruin of an unseasonable fall rain is often total: financially a catastrophe, a humiliation psychologically. This happened to us in 1976 and 1978, when we waited for enough sugar to make enough good raisins to raise enough cash and then got caught with enough rain to ruin the entire crop.

"I ran out of heat," a broken man of seventy remarked to me as I surveyed in October 1982 his 120 acres of stinking, fermenting wet grapes submerged in a sea of muck and worms—for a three-quarters dried grape is still a grape, never a raisin. When given a good inch of rain, it is soon little more than a putrid mass of mold. Years of capital investment were no more than distillery material and z-grade mix for the local cattle feedlot. Why did he, a poor man, pick so late? Why did he go late along with the smug established growers? "I was short with the bank either way," he offers.

Is it then wiser for the tenuous and less secure to play conservatively, to cut their grapes much earlier, in the second or third week of

August? It is. This early picking for those who must have some recompense for the bank greatly increases the chance that the raisin crop will be harvested. Grapes will dry on the sizzling summer tray and dry quickly, given the hundred-degree searing heat of the August dog days.

Here, as every farmer knows, is the trade-off. Early picked grapes are often immature. Without much sugar they scorch in the sun. They fail raisin quality tests. They weigh little. They make a red, not a purple, raisin. They bring in *some* cash but often far less than is required to pay the year's expenses. They cut down the size of the industry's crop for the benefit of the later, more financially secure pickers. Some farmers, then, always prefer the chance of September mush to the certainty of August rocks. "You'll need a tarp on your semi to hold those Wheaties from blowing plum off," a bothersome farmer once scoffed when, stung by last season's late harvest and subsequent total loss to rain, we in our twenties picked too early on August 19. The grapes, without much sugar that year, turned into little more than weathered flakes, not really raisins at all. Fifty to seventy tons evaporated away ("Your kid's new clothes just dried up too," the old agrarian pest laughed). A fourth of the grapes that year were immature. Sour in August, they were little more than water and skins.

The perfect equilibrium, the metron of the early Greeks, is what the farmer seeks. He must learn, if he is to survive, that the balance between good production and reasonable safety is *never* ascertainable until after the fact. During the harvest and the ensuing wait as the grapes dry, excellence is fickle. The "right" decision is always fluid. It changes hourly with the weather. A rain scare on a September afternoon, a storm that threatens to annihilate absolutely every tray of grapes on the ground, turns the early harvester, who picked in August and has his now-dried, poor-grade raisins in the barn, into a genius, and the late gambler with green, sweet grapes, vulnerable on the dirt, into an abject greedy fool. All this will be confirmed by the night's downpour or be erased from the mind in hours as a strong wind—without any explanation or inherent logic, much less "fairness"—blows away the threatening clouds.

In the clear, cloudless heat of morning, yesterday's wise man (who picked early, his crop now dry in the barn) is seen as panicky. Given the new forecast of days of good drying weather ahead, he *appears* to have harvested immature grapes needlessly early and thus to have given up for no good purpose tons of would-be raisins. He is, in short, a hothead who bolted too soon, scorned by the late harvester, a lucky man who is now suddenly transformed within twenty-four hours from greedy and reckless into a paragon of wisdom and steely nerve.

These images, these thoughts, these realities—for they are, in the last analysis, like the weather, real enough—change, day in and day out, within an agricultural community, until the harvest is finally played out, the whole daily, hourly process repeated silently, often unobserved by others in the family, pounding in a farmer's skull, year after year.

A big rain in late August or early September, a farm-ending rain, is more the exception than the rule. It is a frightening event, but one that is supposed to come every ten to twenty years, one whose terror of approach, rather than actual downpour, wears down the farmer. More often the crop is harvested, dried, and trucked in, insidiously each year driving down the price as reserve stores pile up. The real problem for raisin farmers—for all farmers in America now—is never the inability to grow food, but always the bounty and the falling price driven down by the ability to do so. The raisin farmer fears the clouds, but their absence more often destroys him. "Is the price going to be good this year?" we used to ask our grandfather. He'd shake his head, "Go take a drive over to the raisin packers. Their stacks of last year's raisins are sky high."

The San Joaquin Valley summers average well above ninety-five degrees, with little relief at night, steady seventies and eighties even in the dark—twenty-four hours of good cooking weather. Ceaseless radiance ensures Thompson grapes reach maturity by late August or early September. Far more importantly, it is, as I said, not supposed to rain around Fresno until November. High pressure keeps out Mexican hurricanes from the south and feeble Arctic moisture from Alaska. Theoretically the grapes dry quickly once poured on the ground, where there is no humidity, only heat.

Anywhere else in the United States cloudlessness in September is a ludicrous idea. The Southwest sucks in tropical moisture from Mexico and Baja; Oregon and Washington are buffeted by occasional Alaskan storms. Summer thundershowers and hurricanes systematically blow over the South and Southeast. The humid Midwest and Northeast lack the necessary intense heat; floods are just as likely there as weeks of calm. You cannot count on summer in Idaho or Montana. The only region *in all of America* with enough temperature to produce naturally dried raisins for three weeks in late August and September is the Central Valley of California. Men of the nineteenth century went broke, toiled for no good reason, died before they discovered that the ground beneath them was not just excellent grain soil, not merely the ideal home for figs and peaches, but the *only* region in all of America for raisins. Half the world's supply is their legacy. Alma is aptly dubbed the Raisin Capital of the World, for it really is; the chamber of commerce for once is not lying.

True, the San Joaquin is now an increasingly ugly place in the summer, logically disdained by the wiser and more refined coastal Californians, the cheap escape hatch for thousands of scared white suburban refugees from riot-anxious Los Angeles; at the same time, ironically, the favored haunt of the newly arrived from Mexico. But who cares what all these shiftless do in our once-agrarian towns? Let them buy, sell, build, destroy, birth, and murder. Their destruction of valley habitat has not quite yet reached climatic or topographical proportions. So thank God that the raisin farmers' two-hundred-mile depression is still little more than a large dusty frying pan, an asthmatic's nightmare, from the air a stale bowl surrounded by mountains on the north, south, east, and west that stand guard against the entry of all relieving storms and marine breezes.

Raisin farmers do not want those refreshing winds. They have no patience with a cumulus sky that blocks unpleasant sunlight. Much less is there ever a desire for the refreshment brought on by summer sprinkles and showers. They welcome instead only blistering, dry, even dirty torridity. The 103-degree August day, the transplanted weatherman reports to the million of greater Fresno, is terrible, unbearable, and insufferable.

Four thousand raisin farmers disagree. "Whatta day, boys," my bedraggled crimson-faced grandfather used to beam through his sweat at ten on such mornings; "it's hot as a firecracker." Even now put them on the East Coast, Florida, or Hawaii for an early August preharvest vacation, and these odd raisin men can never enjoy a summer downpour, though they be thousands of miles from home. They get edgy wherever, whenever they feel hot water in the late summer sky. They don't sleep when they hear drops on the window.

Raisin grapes, then, need heat, moistureless heat. At the same time, the lush Thompson vine also demands water, lots of it. Those two natural polarities, fire and liquid, rarely exist in unison. The same climate that produces the scorching temperature and the absence of humidity also rarely dispenses winter or spring rain. In short, you need a desert right next to a lake.

Fortunately, unlike the corporate barons on the west side, we small farmers are at the feet of the Sierra, the raisin-sustaining, valley-sustaining Sierra. Most vineyards on the eastern side of the San Joaquin Valley are less than forty miles from the mountains. Our Sierra Nevada is high and steep on the proper side, its western slope right above us. Fifteen feet of snow and more can fall in an average year, melt on the peaks, pour right off into reservoirs, and then be channeled down into communal ditches, watering the vines, and in the process recharging the valley aquifer below. Our farms are the Sierra's primeval flood plain, the runoff now dammed, compromised, and carefully allotted over its former haunts, the valley floor that sits atop millennia of stored flood water beneath.

The old generation of farmers occasionally visit the peaks, not to fish, not to hunt game, not to ski on the lakes, not even to backpack, ride, or sightsee, but to inspect the snowpack, to gaze at their water. The Sierra is the agrarian's benefactor, his lifeline, and I have never met a raisin farmer yet who did not speak of it in reverential, near-religious tones. Even when it played them false in the six-year drought of the late eighties, farmers knew that the snow would return before their ground pumps went dry, knew that their water would pour again down from the mountains, as it had for the past century, before their vines expired. Before the Fresno smog the valley was an Alpine-like recess, a pristine flood basin for the mountains where each dawn the farmer's face glowed orange as the sun came over the snow-melting peaks to the east.

The successful raisin grower below hogs his summer ditch, claws his neighbors for all the mountain water he can get, and hates to turn on his pump. He has no wish to tap his safety net, the subterranean aquifer below the surface, to pay the power company to pump up mechanically what should flow down naturally. In most years the raisin farmer can water his entire vineyard for the price of his irrigation taxes, for the maintenance of the canals his ancestors long dead once dug. Only in a few valleys in Greece, Turkey, South Africa, Australia, and Chile—raisin areas all—do such high, snow-laden mountains tower over parched plains. Their success or failure in any given year shapes the price all others receive who work in like terrain. When the disastrous 1976 tropical storms destroyed 80 percent of the California crop, valley packers searched the world over for raisins. Farmers here claimed that the local cooperative president, Larry Black, even went to Afghanistan looking for what his members had lost. But Larry Black, as we shall see, would do about anything to sell raisins except make a profit for his own bedraggled growers.

Besides the weather, there are good workers in the valley, the second ingredient to the farmer's success. A worker, a young one with a strong back who is under thirty, can pick three to four hundred grape trays a day, a third of an acre himself. I used to have strong arms and a back of sorts, but the most I could ever pick in my preteens was two hundred. I wasn't hungry enough, and I was a little too weak and skinny to fit under the vine and absorb the punishment. Or maybe I didn't have a woman of the type eager to spread the grapes on the trays for me. I did learn, though, that it is the worst job in America, a vocation right out of the poet's inferno. Nothing in my own agricultural or academic experience—dusting vines in a fog of Dibrom toxicity, periodically climbing down into the century-old cesspool to shovel the grease off the sandy bottom, packing bearings inside a chemical spray tank, or trying to teach Euripides and Catullus each year to the distracted and contemptuous illiterate at Fresno State—comes close. So the harvesting of raisins has been the profession of a succession of the valley's dispossessed, first the Armenians, Germans, and Swedes, then the Orientals, the Okies, and for the last thirty years, almost exclusively the Mexicans.

I say Mexicans, not Mexican-Americans, because grape pickers, thirty thousand all, are almost now all green-card holders. Whether real legal card holders or illegal fake card holders is hard even for the most scrupulous to tell from the documentation presented. Most illegals carry more identification papers than do wealthy whites. Ask for the mandatory I-9 immigration form, and social security cards, driver licenses, passports, charge cards, zoo memberships, and car registrations pour out as well—the majority of them expertly forged, in better shape than those in the wallets of most Americans.

One generation in America, and few Mexicans work any longer in the fields. To the right-wing self-made in farming, the border crossers are corrupted by welfare and lose their edge, preferring sloth and criminality to the nobility of incessant toil in the agrarians' vineyard. Self-esteem and ethnic pride are sometimes found by their offspring in lethal gangs that kill and maim one another, while culture is forfeited through unendangered work in the fields for the non-Hispanic other. Not a few farmers have noticed that a missing roll of toilet paper in their porto-toilets makes them clear and present dangers to the health of the entire Latino community, while the flannel-shirted, floppy-panted, hairnetted *vatos* in Fresno now gun down their own kindred, dumping them gutted in rural peach orchards and irrigation canals without much rebuke from the community politician, labor organizer, or parish priest. No, the vineyard does not slay its workers, but it can be the scene of their final disposal. For the real Latino killing ground go instead to the crossing of Martin Luther King Boulevard and Cesar Chavez Avenue in Fresno, not beneath the sweltering vines amid the spiders, dirt, and grapes.

The urban left-wing dupe reasons otherwise. Farm workers finally refuse to be exploited and so say no to work at poor wages without benefits. The permanent Mexican alien purportedly will eventually move off welfare to more ennobling professions like hotel, restaurant, and garment employment or lawn service for the affluent. The Bay area professional may still boycott fresh grapes (but why not wine?), but he thinks nothing of the servitude of his Hispanic gardener, maid, waiter, or dishwasher. Forget the minimum wages, the absence of payroll deductions, the payments in cash. Has he not, the urban progressive reasons, rescued the underprivileged from the clutches of agriculture?

Where the truth lies is paradoxical. No grape picker can sustain himself in America as an American on farm-labor wages. There are no deeds to property, no day-care centers, vacations, VCRs, big-screen televisions, meals out, jet skis, life insurance, and deferred income on five or six dollars an hour, three to four days a week in the vineyard. You really cannot in a generation take a Third World native from the tropics and turn him into a bumbling suburbanite wrestling with his weed eater.

On the other hand, should the alien farm laborer live ten to twelve to an apartment, should he rotate beds, pick up an aging two hundred-dollar gas guzzler, pay no car registration and purchase no car insurance, have no paid driver's license, genuine social security number, or medical insurance, and give no heed to the summons in the mail, he can live like an un-American in America. His trash, his broken radio, his moldy mattress? Along with his litter of unwanted three-day old Chihuahuas, he throws them all out the truck at night into your vineyard. He survives as a metic, a resident alien in a foreign *polis*, excusing himself from the liability and responsibility that owning even a modicum of property in America now entails.

Run a stop sign outside of Fresno? You yourself run, leaving car, wreckage, bottles, and bodies behind. A vendetta over a woman? You stab deep and flee south. Become sick or pregnant? You wait until your final hour of crisis arrives and then enter the sophisticated and absurd world of American emergency room medicine to find either salvation or a sanitized death. And lose your mind? The local rest home in town is full of demented farmers and vineyard laborers alike, side by side, bed by bed. It makes little difference to the nurse's aides whether you're brown and dribbling or white and incontinent, whether you're the veteran Zapotec picker Hilario Montoya on Medi-Cal or old Elbert Fall drawing the final bit of equity out of his eighty-acre vineyard mortgaged to meet his last tab. The money, public or private, flows, the chairs get wheeled, the diapers are changed, the bibs stay on; in California the illegal can still get equity in the penultimate twilight before the darkness.

With luck and ingenuity, then, the worker of the fields can salvage far better an existence on the fringes up north than in the center down south. For a while, that is. Only by living in the shadows and ignoring not some but every statute is there hope that the vineyard laborer can save half his wage, in a single year sending three to four thousand dollars in cash back to Mexico—the equivalent of more than three or four years of steady work there. What valley farmers that are left (in their self-righteousness) claim is that *if* grape-workers avoid the blade, *if* they stay out of the way of a bullet's trajectory, *if* they keep away from the alcoholic-laden car, *if* they keep their bodies free from human-induced disease and injury, then by thirty-five a Mexican can recross the border with thousands of American dollars in his pockets.

I don't think that ever happens. Do all that many average Americans take their retirement funds to Mexico to live cheaply in that southern paradise? There are attractions in America beyond clean drinking water, flush toilets, and the local efficient hospital. A South American of twenty-five to thirty, however disciplined, however determined, still needs a family at his side and so must have a life beyond himself, a costly life, as a guest worker. I think most laborers, if not killed by one another, injured, or jailed, stay up north to spawn an impoverished but enormous family that inevitably plugs into the declining social services of an exhausted state. Without money an illegal immigrant can still strut in town if, cometlike, he has attached to his rear a long winding tail of a second litter of offspring. Less often, some return humbly to the predictability of a familiar poverty across the border. I repeat, few, if any, retire with thousands to Mexico.

No one has explained to me yet how these dispossessed come up from Mexico with a stronger work ethic than Americans and often in two or three years have less of it than we do. How can we near instantaneously make a laborer who is better than us worse? Is it, as we suspect, government welfare and the accompanying protectorate of entitlement and subsidy? Or is it MTV and rap, our absence of formidable religion, the collapse of shame, the rise of victimhood, the ubiquity of cheap and unnecessary goods, the nausea of advertised hype? Is not the cause the old crude American drive for pelf and disdain for physical labor? Somehow all the rural Indian's prerequisites for astonishing work—poverty, large families, religious fatalism, machismo, cult of the young male—are themselves the most susceptible to the corrupting enticements of our now bankrupt American culture. We can take any Third World peasant, rob him of his inherent nobility, and then reinvent him with a horrifying evil unmatched by its now agape creators. Spanish folk songs and straw hats give way to rap, earrings, tattoos, and shaved heads in less than a year. For a lazy no-good American white boy of the suburbs to lie in sloth before daytime TV, to blow a few numbers, to poke his thirteen-year-old girlfriend on the couch is but a transitory flirtation with the wild before his perturbed guardians slap him straight on into insurance or sales. For the Oaxacan without such second and third chances, all such acquired American liberality is lethal. Laziness, drugs, and babies confirm that the vineyard leads not up but down, far below into either a nether world of savage criminality or pathetic ethnic chauvinism.

It sizzles in a vine row. Anything metal on your clothing, belt buckle, buttons, jewelry, burns the skin. Like the Bedouin, you try to wear lots of clothes, never shorts and T-shirts. Unlike ourselves, Rhys Burton wore heavy bib overalls, scarf, long underwear, straw hat, gloves, and heavy boots—in July—on the theory that the sun burnt the skin and cooked the brain. Even when it's over a hundred degrees, most pickers have long sleeves and a bandanna around the neck to keep the dirt, the dust, and the juice off the arms and neck and to prevent vine detritus from entering the shirt, lower backside, and crotch. Elbows and knees, if not to be polished raw, need cover from the sand. I've seen stalwart workers with flannel, even woolen shirts when it was over one hundred degrees. After about every fifty trays or so, you run into either yellowjackets or black widow spiders, the venomous ones that hide in the shade under the vine, that react violently when a hand scrapes their silk or paper nest. A picker can also get stung or bitten by the human predator; he can get caught in some sloppy farmer's trashy field, where puncture vines, sand burrs, and Johnson grass strangle the vines and scour your knees and ankles, cut your hands, or leave seeds stuck in your sweat.

These sly agrarian birds cut expenses on their cultivation, hoping that someone desperate enough at the end of the year might be duped into wading in to harvest their mess. The old widow farmer Clairice Moss around the corner despised these sloppy, weedy farmers. Near toothless and wrinkled beyond recognition at eighty-three with her colostomy bag all but disconnected, she used to drive around to spy out the weed patches of her inferior farming brethren. "I hate a dirty vineyard," she scowled out the window to any farmer unfortunate to meet her on her rounds. She made quite a little speech to me once, "Your grandfather wouldn't like you spraying those weeds. You gotta dig them out by the roots, gunnysack them, and burn 'em up in the alleyway." She almost fell over as she got out, demonstrated, and, now odorous and sweaty, hobbled off in disgust. Later I learned that, in fact, Rhys Burton and his siblings had hired themselves out at the turn of the century as Johnson grass diggers, a crew that would dig, bag, and burn others' noxious weeds when there was the occasional need for additional cash.

Her own ranch, after her demise a year later, was foolishly left to urban heirs, bickered over, and sold off. They took to court eighty-year-old Delbert Ridgeway, her nephew and foreman, their own cousin, to pry him out of one of the farm's many small frame houses. He had good reason to hold Clairice to her word, to claim a clapboard house as retirement in return for thirty-five years as an anonymous underling. Clairice may have known Johnson grass, but like many farmers she was hard on her own kind, softer on the prettier, more distantly kin in town. I testified under oath on Delbert's behalf at the eviction suit—an ancient, wrinkled, bald, and toothless tractor driver surrounded by the stylish, manicured and trimmed, urbane relatives who wanted the $25,000 house empty along with their newly acquired 160 acres of vineyard to facilitate a clean, quick sale. I told them that Clairice while in the rest home had told me the house was for Delbert (she had). "But what was her state of mind," screamed the carnivorous attorney.

"Ask yourself, she talked to me about Delbert about the same time you must have redrawn her will." So she had; case dismissed.

Most pickers start at 5:30 A.M. and quit at two in the afternoon, winning their fifty to seventy dollars gross (eighteen to twenty cents a tray times three to four hundred trays). The conquered guys over forty who are about played out usually do not cheat on their row count. You can pretty much depend that the number they write on their first tray is about the number of trays they actually picked in the quarter-mile-long row. They also don't leave grapes all over the ground. Pride prevents them from milking the berries (using the hand to grab, instead of the knife to cut, grapes), spilling infectious juice on the bunches. And they rarely omit the tiny, hard-to-get cluster on the vines. The good pickers who came up in the 1960s and 1970s manage to have all the berries neatly on the paper and leave the trays in a straight line all down the row. Bent double, they stagger out of the vineyard at day's end. They have picked fewer of these excellent trays, and so they have made nearly nothing.

Mostly, the war—and make no mistake about it, it can become all-out combat—is with the young lions under twenty-five, the macho and lean who will be *cholos* in a few years if they stay up North. I do not forget their names. Pugas, who tried to run over my cousin with his Opel station wagon after we fired him; the Chapa brothers, who gave us the evil eye all day and stole anything not tied down (why would they want my four-year-old's rusted scooter?); Thick Lips (Cruz Galindo), who always took off his shirt and threatened with the big knife, promising to cut us if we didn't say that we liked his job. We didn't like Thick Lips' work, and he never tried to slice anyone because he always needed our help getting his car running and out of our vineyard. Pride shrinks when you need transmission fluid. Poverty, Juvenal says, makes one look ridiculous. I'm sure they had names for all of us as we charged down the rows, pulling up trays, pretentiously trying to look formidable, and leaving empty threats of dismissal everywhere, as if those who picked were our real robbers, as if they might by themselves bring us down.

Once in a rare harvest season every farmer and farmworker know it goes beyond mere posturing. During grape picking of 1982 (long before the construction of the circuit wall around the house), two anonymous and intoxicated figures parked in the yard about 2 A.M., at first no more than shadows breaking beer bottles on the pavement but soon boasting in broken English of theft, mayhem, and assault to come to us inside the house. The response time of the beleaguered Fresno county sheriff in late summer is typically about fifty minutes. Why bother? This time it is no bluff, and only a sixteen-gauge, century-old shotgun jammed inside one man's now silent mouth keeps them back. Urine runs down his pants' leg. The real worry was that the antiquated cocked hammer might slip prematurely (as it does often) and so decapitate for no good reason the coward now suddenly slinking, now so cooperative, now so religious. No farmer, however nervous and apprehensive, however eager to enforce the laws of hospitality and good manners, wishes to blow off the head of an obnoxious thug. As much as I would have liked to I did *not*, triumphant in this standoff, like Odysseus to the Cyclops, boast to the cowering criminals, "I am Morgan Hendrickson who humiliated you, son of Axel. I am here twenty-four hours a day; come and get me anytime if you dare." I was no Homeric hero, more a homely Thersites who sat down in the night glad to have it over.

The days of paternalism, private charity, and mutual respect in the vineyard—for good or ill—are far gone; then a Rhys Burton might walk into his vineyard, be addressed by a small band of steady workers with "How are we doing Mr. Burton," and reply with "Now you work steady until three, but get indoors then, you'll sunstroke out here any later than that." Myths still abound about the Japanese pickers at the turn of the century, the Filipinos in the twenties, the Okies in the thirties and forties, and the braceros bused up from Mexico in the 1950s, the golden days of agricultural labor and petty exploitation, where the gap between small farmer and day laborer was not wide. The myths abound about their industriousness, their perfect raisin trays, their politeness, and the reciprocal respect between farmer and farm laborers, when their children and we grew up together on the same ranch, went to the same schools, and had the same friends. The union, the government, and the immigration explosion in Mexico put an end to all that and so created the myth of the golden age of grape picking that never really was.

These younger, more aggressive, and sometimes criminal workers are now serious moneymakers, and so they pick both fast and sloppily. Their grapes are everywhere. Their line of trays looks like a snake; papers are either put too low or too high on the terrace. Long after they're gone, the corners of our trays then get shaded out by the top of the vine and never dry. Nothing is worse than to have a green bunch in the raisin roll. It can rot the whole biscuit.

Usually these fast pickers pick a light, a disgraceful grape tray, a few bunches piled in the middle once they get deep into the quarter-mile vineyard and you can't see what they're doing. Then a week later, when it gets hot, you are left with a few individual berries per tray that burn up. Themselves long gone, their damage remains to plague you, a bitter remembrance of their unpleasantness. The only adequate job these fast pickers do is on the row ends, expecting that you won't walk all the long rows and check up on them or that you'll be too timid to say much if you do wade out into the middle of fifty men of their stripe and find a mess. The young and combative pickers do a terrible job. They cheat you and thus make more money. They add fifty or a hundred trays to their count, overcharging you nearly ten or twenty dollars a day. You must walk the row! Your own kids must walk and recount their thousands of trays, checking always their fraudulent arithmetic! Forget about the pretentious gifted program at the local school, which is neither gifted nor a program. It is far better for a child of ten to walk among swarthy, callused men of the knife and the bottle, checking their veracity with a cheap hand clicker.

I say a tough few are terrible workers, but it's a relative term, based on whether you spend or do not spend August and September on your hands and knees. From the grape picker's point of view, any farmer who can walk upright has reached success. In their eyes no farmer or his family could last more than a few days on his knees in his own vineyard. Their arms alone transfer *all* the grower's year's investment from vine to tray. To them, the ramrod back of the farmer, his steely visage, is a sham, counting little unless he is on his knees.

My grandfather's brother, Blanton, once in the 1930s scoffed at the expense of their labor. In the midst of the depression, he supposedly reasoned he had no need of such costly grape pickers, no need of a bank loan to pay someone else. Blanton put his wife and kids out in the vines to pick every day from August to October. For their inevitable failed heroics and lost crop, the local bank soon foreclosed and asked them all to leave. For the last sixty years our more sober line has farmed that more productive sixty acres across the street. We grew up in the house that my dad moved on to Blanton's old farm, where we were reminded of the value of hired labor.

In fairness to the farmer, farm toilers at least get paid in real money, our money. No sane grower stiffs his workers. The loss of repute and bodily safety curbs that pretty well. In answer to the workers, we have the excuse that we may not get anything for an entire year's work. We'd rather have bosses like ourselves writing our own checks than the fellows who run the raisin industry. An empty suit at the local raisin cooperative can coolly announce that one's entire capital reserve—a family's $40,000 retirement—is simply gone. In 1990 our local packer wrote us to bring in the raisins. The rumors were false, he said, the line of credit secure, the payment ready on delivery. But the thievish operator of the plant, a corporate puppet laden with merger and buyout debt, had forgotten about his own obscure Taro Kobashi, perhaps the last honest field man in America, who tipped us off, and so they never got our raisins. The poor credulous farmers whom Taro could not reach delivered and got their harvest checks tied up in a bureaucratic knot.

Do farmers ever slay that embezzler, the coiffured advertiser, the fragrant broker, the high-salaried parasite? They do not. They do not follow him home to his North Fresno estate and burst through his double oak, stained-glass doors to slide a blade in his ribs. The snookered do not seek to carve up the suit's accumulated pile, his house, his car, much less paw his stylish wife. More often, the worst among us *admire* the white-collar con man, the cunning and the sheer braggadocio involved in explaining to thousands of impoverished agrarians that men of his ilk have concocted a way to rob them.

Is not the modern suit a special species of hydra? Cut off a head, and he grows another one, legal, with a longer, shinier fang. Burn appendage two off, and it reemerges from bankruptcy court fully toothed and commissioned in a new office. All that is needed is a new phone and desk, and the *vipera vitis* goes to work. Blinkered grape farmers never even take a stab at decapitating that beast. We would rather go after what we can see and understand, the light tray, the impoverished man who leaves a bunch or two unpicked on our row.

Every farmer must concede that anyone who does vineyard work is one up on *any* American, white, black, yellow, or brown. None of our citizens can last a day or two of raisin harvest. Usually, when the picking starts, a carload of poor deranged whites, blacks, Mexican-Americans, or Hmong on welfare drive by the vineyard and say they are "directed" to work cutting grapes. So much for the local "human services" office's most recent pathetic efforts at "workfare." Apparently, grape season at least gets the lethal mass of slackers out of their offices for a day. Tattoos, guns, bandannas, knives, drug-induced malnourishment, wraparound sunglasses, tooth-gapped death stares, and Oakland Raider hats—the whole visual apparatus of the American criminal class that so frightens the rest of the world—do not amount to much when you're alone, crawling your way through two hundred vines in a row a quarter-mile long.

Rosalio Hernandez, a five-foot-two-inch, one-hundred-twenty-pound native of the Yucatan, who traveled one thousand five hundred miles northward to pick grapes, whizzes on by, a human dust cloud, the grape trays flying down on the terrace thirty, forty, fifty, and more an hour. Pickers born in America fare not so well. A quarter-mile back in the vineyard, frightening ebony Trei Hodge, bulging with prison-spawned muscles, surly Ric Lara, third-generation unemployed leech, and gaunt Eddie Rasmussen, the emaciated local fence, sweat, curse, threaten, and then storm out of the field after twenty minutes, unconcerned with recompense for their $3.40 worth of trays picked, glad only to be free of the dirt, heat, and humiliation of hard, honest work. Anyone who picks grapes, no matter how much he cheats, no matter how poor a job he devises, is a harder, tougher worker than any American, farmers included.

And finally, you wonder, where is the union, the United Farm Workers, who drew in the Kennedys, the movie stars, the priests, and professors to places like Delano and Tulare? For much of the 1970s its membership was on the rise; conditions in the field improved; real wages of farm laborers soared as the country roared on with inflation. Urban liberalism, too, found the picker and pruner chic, romantic toilers all, and so boycotted fresh grapes of every species in the stores, oblivious that those same hands had also harvested the wine, peaches, and raisins in their shopping bags. By 1975 many around Mr. Chavez were prophesying an AFL-CIO–like paradise in agriculture: journeyman grape pickers; class II forklift drivers; apprentice plum thinners, step-five grape tray layers, all overseen by a complex hierarchy of grandees and fiefdoms—organizers, hiring halls, monthly tribute deductions channeled into pension funds, political action committees, lobbyists and lawyers, all clinging to the carefully constructed weather-beaten visage of Mr. Chavez, the man with the hoe. Understandably, the union welcomed the corporatization of California agriculture: all the better to organize, all the more capital to be tapped, all the easier to depict farmers as rich, white, and rapacious. One big farm, one big union, no different than autos or steel, clear-cut good versus age-old evil.

In contrast, the rugged and independent yeoman of twenty and forty acres was often a bothersome obstruction: many had developed mutually beneficial "paternal" relationships with generations of nearby laboring families; many were not much better off than the laborers themselves, living among, going to school with, working alongside labor. And unfortunately for the union, many of these adversaries were not Anglos but themselves of dark hue and strange tongue, no strangers to prejudice and adversity. It is difficult to pin racism and agricultural exploitation on a Sat Abe or Sam Mukai, staunch members of the local antiunion Nisei farmers coalition, thrifty residents of small cinder-block houses, custodians of forty-acre raisin vineyards, themselves veterans of land confiscation, deportation, and the internment camp, no strangers to "Jap," "Nip," and "Gook," often dubbed "Chinos" by Hispanics themselves.

Equally perverse to package as exploiter and oppressor was the growing tide of no-nonsense, union-hating Punjabi immigrants, who pooled their scarce resources, formed lending associations, and purchased money-losing raisin ranches; tough and intrepid men like Kewal Singh and Sukhvindeer Istanbouli, jet-black, turbaned, sworded, and robed, to be dismissed cruelly and with prejudice as ragheads and cow jockeys by an ever-growing number of envious detractors. So the family farmer—Pakistani, Armenian, Japanese, Sikh, Portuguese, Greek, Mennonite German, Scandinavian Lutheran, and Mexican-American—was a nuisance to Mr. Chavez as he had been a thorn to agribusiness moguls and bureaucrats for decades, this oddity who could not be pigeonholed by class, gender, or color, by anything other than his peculiar desire to be left alone among his trees and vines, an appetite that leaped across race and found its way into the hearts of strange but kindred men.

Still, the union business would all have worked, I suppose, but for the bizarre political coalescence of left and right who, cheek by jowl, conspired in the 1980s to open the floodgates of immigration and inundate California with millions of cheap laborers from Mexico. To the reactionary businessman of the 1980s, this wave of imported workers was the natural expression of laissez-faire capitalism: the market alone would adjudicate, sucking in available labor on the North American continent as it saw fit; alone unhindered by artificial and noneconomic borders, the shop alone most efficiently might allot workers to the myriad tasks at hand in the new "global" marketplace. Better then to allow the meddlesome Immigration Service to die on the vine; better yet to have its intrusive agents exit the workplace altogether, "big government" bureaucrats all, who had no grasp of the organic ebb and flow of human capital. Such conservatives, such patriots, these captains of industry and agribusiness!

On the other extreme, the atonement brigades and La Raza demagogues just as eagerly welcomed this immigrant bonanza, this radical peopling of Alta California. "The border crossed us, not us the border," they screamed on the evening news. Somehow they were convinced that soon these millions of undocumented and illegal workers might find citizenship, affluence, and education: a new majority of mestizo suburbanites who could maintain cultural purity and so quite wisely elect refined and sophisticated lighter-skinned representatives such as themselves. Quite unconcerned were these magnificoes about the short-term destruction of wages among their citizen brethren in the union. Government entitlement—and the representation to ensure that subsidy—could suffice. They also had no idea that Mexicans were running, wading, and swimming to escape, not embrace, Latino culture, fleeing to the bounty that accrued from the bitter discipline and self-reliance of Anglo-Saxon tradition and Western custom: the free economy, constitutional law, and a social contract bequeathed by skeptical and worldly men like Hume, Burke, and Adam Smith. For all the bluster born of insecurity and inferiority—the Cinco de Mayo marches, flag decals on windshields, and block-lettered La Raza graffiti—few Central American Indians had any desire in their hearts to re-create up north the fiasco they had just escaped down south.

And so after 1980 the people were tacitly welcomed to El Norte as never before. Accordingly, compensation in hotels, restaurants, gardens—and agriculture—froze and sometimes plummeted. Bodies outnumbered the tasks at hand, as strong backs from the South (without union cards) outlifted, outpicked, and outpacked softer American hands. Mr. Chavez's brotherhood of agricultural toilers vanished root and branch, its fate sealed for good. And as if by divine sanction, the economy soured at the same time. Liberalism dissipated, backlash grew. Those who had once boycotted grapes now worried about the graffiti on their park benches and gunfire in their upscale schools. Dusty Indians from Central America dotted El Camino Real in Atherton and Menlo Park, hoping to be picked up for yard work by computer programmers, as if they too would sip wine rather than guzzle beer, as if they would not urinate in their park hostelry at night, as if they should take the red-eye home to Yucatan each evening.

The subsidized pigmentists who once grandly called for a Hispanic Nation, brown power, La Raza Unidada, who objected to public support for the local zoo as a white subsidy at the expense of bilingual radio, now in the sea of unpopular immigration scurried for cover, terrified at the reactionary white monster they themselves had wakened and energized from his stupor, a foul and odious specter they alone had coaxed and conjured back from the dead. "It's a white thing, you wouldn't understand," I saw on an angry kid's t-shirt in town—the White Race now in turn finding support from the decades-old legitimacy of La Raza. How pathetic that in California pink Okies and swarthy Greeks were to seek (like the absurd inclusive notion Hispanic) a monolithic unity in their pigmentation. Anglo strangers in Fresno stores or at rural Little League games now felt a bizarre kinship, swapping advice about preferred handguns or horrific accounts of break-ins, burglaries, and assaults. The United Race was not going to be so bad an idea after all. At all this the bewildered farmer simply grimaced, glad enough to have trees or vines rather than people in his midst, sorry only that his forty acres could not be airlifted out of California altogether.

Meanwhile, the market economists suddenly worried that the pool of idle bodies was not so fluid after all and that unforeseen laws—other than supply and demand—might govern whether a family returned to Mexico or stayed on the dole in Fresno when the jobs evaporated. So the wages of the agricultural worker did not rise, and like the farmer, the farm laborer was left behind in the debris of the eighties.

But you must forget about race, the union, the pundits, and the provocateurs on the right and the left; you need only remember that the farm laborer is like the family farmer—inherently a noble creature who, crushed daily, crawls out on the walk again to be stepped on, smashed, and ridiculed, oblivious to all that would eat his fruits. I care little that the demagogues tell me the borders crossed them, not them the borders, as if there was a single Hispanic in the empty valley in 1880, as if there was anybody of any color who wanted to camp out on this 120 acres and grow raisins beside that desolate tule pond that not even the railroad wanted.

I think only of individuals, not causes, theories, or castes. On a September Friday in 1984, Javier Moya and his family of nine pocketed their wages for a week's picking, eager to break off at two for an afternoon picnic at the Fresno zoo. By 2:30 they were no farther from the vineyard than the dusty alleyway, the power-steering pump in the station wagon burned out. Javier Moya and his oldest two sons have only a general idea of hydraulics but a very clear belief in not being towed into town for a five-hundred-dollar repair bill. At four, three trips to town later, two hundred dollars poorer, Mr. Moya has his replacement pump in his hands, but now takes a break from under the car to hitch his three-year-old back into Alma for a shot of penicillin for a neglected strep throat. Roaring out of the vineyard by six, he is abruptly thwarted at the pavement when, *mirabile dictu*, the master cylinder fails. Another journey to town, another two-hundred-dollar contraption, and the family dines on stale tacos as Javier plunges into the dirt beneath the car for round two with this other species of strange pressurized fluid. By dusk, with steering but no brakes, his week's pay nearly half gone, still dirty, he asks that they spend the night in the nearby orchard to save the gas of a needless trip home. It is better anyway, he smiles, to work on Saturday morning, and that is a mere nine hours away.

Besides the excellence of the climate, the plenitude of water, and the superiority of the labor, the raisin farmers themselves are first rate at what they do. I went to school for seven years at Santa Cruz and Stanford and for two in Greece and have now lost another seven years as a professor at Fresno, two more years as a research fellow and visitor at Stanford. None of the scholarly grandees I met during those two overlong decades—and many were talented in a variety of disturbing ways—were as bright, as wily, much less as resourceful and ingenious, as farmers in our environs.

Out of politeness, I omit similar comparisons of respective toughness, courage, resistance to pain, and honesty involved in contemporary academic and agrarian life. But anyone who took one look at Iver Johanson wading out of a muddy vineyard in his irrigation boots (despite his enormous belly and square pink head), at one-armed Ray Mix perilously close to losing his remaining appendage to yet another wobbling PTO shaft, or at cigar-eating Jake Norbanian with considerable proboscis, razor-sharp pruning shears, and dirty gloves, thousands of numbered expenditures and profit margins automatically being tabulated each second in his brain, and then ran into the polo-shirted Leslie Doily prancing in the classics library at the university or has been strapped down in a two-hour Greek paleography seminar with a high-priced European free agent like endomorphic Professor Rheinhold Botkin at the wheel, would know instinctively what I'm talking about. I pass on so many others, like the pompous English classics import Graham Onians and his sudden fascination at seventy with his own lifelong sexual ambiguity and thus the much-needed supporting apparatus of Michel Foucault. For these great thinkers with elbow patches on their corduroy coats and for the lesser brood who now quote Derrida but snicker about the noble Camus, a broken radiator hose, a cracked window at home, or a leaky hot-water heater results in near incapacity, an embarrassed, distasteful brush with the elements and the call to unpleasant men who must now master those elements. The *apparatus criticus* to Aristophanes, the rhetoric of manhood, the poetics of gender, all that is no help at all in keeping the water in your tub warm.

Most of these raisin growers—like Iver, Mix, or Jake—that are still left, who haven't died, sold, or rented out, are clannish ethnics, men and women who really have not much in common with the rest of American culture—complacent and soft white America or now its whiny and vocal professional minority. Most city dwellers are surprised at that. The idea that a high-pitched screaming Japanese or sullen pitch-black Sikh can, just as easily, just as frequently as an old rose-colored and wrinkled Dutchman or Swede, be hell on unreliable Mexican workers, laugh at environmentally sensible legislation, ignore the entire ideal of a communal irrigation ditch, goes against everything they want to believe about the Third World solidarity of the oppressed. I suggest that there is no rainbow coalition of non-white peoples, at least in agriculture, no real color guide to the exploited and exploiting, no longer a white monopoly on prejudice and gratuitous slander.

Amil Betinia, the near-black Gypsy fruit peddler up on the corner, who ribbed me about the Mexican-Americans married and born in my own mongrel family, once remarked that seventy-year-old Vaughn Kaldarian was very lucky to have such a pretty woman accompany him in his truck. Yes, I agreed that he was proud of his half-Armenian daughter, then an attractive woman of thirty. "No, you fool," he yelled, dissolving my genuine tributes to her beauty, "not that skinny ugly brown girl. I mean his wife." "But she's a Hungarian blimp, sixty, and 250 pounds," I scoffed. Betinia in disdain drew up his wrinkled toothless face, "But boy, she's big and white, boy, white as snow."

These agrarian holdouts, the Betinias, Kapoors, Kaldarians, Kimuras, Avedesians, Singhs, Brahrs, Soareses, Van der Wirts, and Johanssons of the valley, who have no other income than their farms, whose success or failure depends entirely on their muscular strength, nerve, and sheer cunning, are more than mechanics, welders, and agronomists. They are often unrivaled mathematicians, sophisticated accountants who have every penny marked out in their income and expense ledgers. Their entire universe is the constant tabulation of the cost to produce a ton of raisins and the final price they will get for it two years later. Lose a dollar wrench? They hunt for it for hours. Their living expenses? They don't figure it in at all. What they need to run their places for a year, that's all they ask from the bank. For themselves and their families, they eat from what they can squeeze out of their place or their neighbor. If it's a sixth spring sulfuring or a family outing to the mall, the money goes to the vines.

For purposes of social intercourse or simple conviviality, these vanishing holdouts are a distasteful bunch, if anyone who works twelve to fourteen hours daily can ever be dismissed as distasteful. If they save six hundred dollars an irrigation by not pumping, they'll skim off your turn on the ditch. Vaughn Kaldarian stole our water from 10 P.M. to 5 A.M. every day of our one-week turn. Despite verbal threats, padlocked gates, and evening patrols, the nocturnal phantom somehow reached the water. Always the water he sought. It was not the mere savings, it was the water itself, the water he wanted.

In hard times the hard-nosed few are ready. I even think they prefer the challenge of depression to the calm of bounty and affluence. They relish the dare to take their losses out of the fat of their ranch, their workers, their equipment, their family, themselves. All the agrarian others, with the exception of the inherited wealthy—I mean now the lawful farmers, the near organics, the sticklers for paying their workman's comp and social security, for reporting their cash sales, for having all their toilets clean and ready, for registering their underground fuel tanks, for following all the rules of pesticide application and disposal—they are reeling or long gone. They are no more than farming's vestigial organ that hangs limp, dragging the dust, still somehow attached by the torn skin but with no real function, no role any longer to play.

Extinct, I know now, is this yeoman out on the tractor, his dutiful wife writing the checks on the kitchen table, children on ladders in the orchard, all harmoniously producing food in order to produce offspring who will not cuddle and snuggle for an extra dollar. Good riddance to all that, I have heard the poststructuralist philosopher on the coast exclaim, the tweedy radical who locates the American farm as the criminal spawning ground of homophobia, sexism, racism, and capitalism. The Masonic temple, the evening grange discussions, the

Wednesday chat 'n' sew club, the ladies' literature teas, the farm bureau banquet, the Monday night improvement club, the entire evening expression of agrarian culture, I learned at the universities on the coast, had, in fact, been exporting hierarchies all along that poisoned our citizenry. My parents' and grandparents' evenings were not to build and cement agrarian communitarianism, not to improve mind and soul. No, erudite professors told me, such rural traditions and pedestrian organizations without prestige or hierarchies were still delusions designed to cram oppression down the throats of the "other," every species of oppression known—the nuclear family, the unforgiving Protestant work ethic, patriarchal religion, and the racist chauvinism of a mindless and laboring heterosexual middle class.

So now that yeomanry and its cargo have passed, so now there are no more odd Willems and Schuylers reclaiming scrub beside tule ponds; by God we at long last graze in the pastures of their promised utopia. I know what the twenty-first century shall say of *agricolae exstincti*. But what pray will historians make of this other more curious late twentieth-century species, this whiny lamprey who slithers amid the swamp of American materialism only to turn back out of the muck to stick his tiny fangs into his bloated mother, now pouty over all the ingested pabulum that has made him fat but colicky?

But the hard-nosed of all races, genders, and ethnicities are now the tiny muscles of the industry, a farming race alone who have no belly for clubs and get-togethers. Alone they compete with the corporations and the insurance companies. They need no affirmative action, for they affirm daily in deed and in fact that they are tougher and, if need be, more vile than the corporation itself. On the other end of the farming vise, they still can vie with the weekend hobbyhorse farmer, the prissy urban fugitive with money in the bank and a job in town, the high school principal who shops for cabbed and air-conditioned tractors on the way home from work so he can play on his twenty acres. The hard-nosed survivor who lives and works his own ground with his family drops the gauntlet and says, "Out produce me, spend less than I do, see if you can take my ranch, for surely I will do all that to you."

"I'm glad you're painting your grandfather's house," Mohinder Kapoor told me shortly before his heart medicine finally burned out his kidneys. "I like old two-story houses, so it will be nice for me when I move in in a year." Mohinder's son, of a lesser, a more human breed, never moved in. He was run off his dad's place by his tougher sister, who alone inherited the Kapoor gene. All he ended up with for his years on the tractor, for the growth on his neck, was a small house down the road with a mortgaged twenty acres. Mohinder would have approved.

As we fifth-generationers slide into gradual but inevitable financial ruin, we who were once products of the comfortable middle agrarian class, who were taught that our land is a wonderful paradise that never turns on those who would just work, we wonder now hourly, obsessively. We discuss after the raisin crash whether our continued compliance with rule and regulation, with a belief in the dignity of labor and the concomitant need to pay over the minimum wage even in a money-losing year, whether our self-appointed and conceited notions of goodness are what killed us, whether we were too unimaginative in the arts of agrarian fratricide, of "eating your neighbor's shorts," as they say, too silly, too polluted with college educations, too eager to be obedient citizens of some humane and imaginary *polis*.

Just as frequently, I confess, we acknowledge that in some sense we were simply inadequate; and so perhaps Shakespeare was right when he said that the weakest fruit drops earliest to the ground. Afraid or unwilling we were to wade into the dark realm of the hard-nosed, to embrace the kingdom of cruelty and lawlessness necessary for survival now in rural America, our polite bellies lacking the fire of the hard-nosed who will say and do anything, and more than anything, to feed their family and keep their land. For them, stealing ditch water or spraying stolen or unregistered chemicals is an afternoon's amusement. For them, to stiff the impoverished brush shredder or pay the trucker eighty cents on the dollar is standard custom. For them, to pack green fruit or moisten down their raisins for added weight is the canon of success to be passed onto children. For them, to tell their tractor driver of twenty years in a second, "You're all through," is the price of agrarianism. For them, to put their children out as near–indentured servants in the field is the tab for country living. The future—if for any there be a future in family-farming in America—is surely theirs, not ours of an inferior mark, who with a self-righteous sob and a yelp will shamelessly lose our land, our families, ourselves, all for the conceited avoidance of the talons and the curved beak.

Are you fearful of the sight of a huge and towering telecommunications pole in your vineyard, its microwave rays bouncing throughout your orchard, its silhouette an insult to Willem Burton's hundred-year vineyard below? Don't worry; the hard-nosed will welcome the company over to plant that monstrosity among their vines at five hundred a month rent in perpetuity. Concerned that the widow of Augustino Zavala is now seventy-four and useless, with nowhere to go other than the small frame house beside yours where she has dwelled as your foreman's wife for thirty years? Evict her and rent it out at three hundred a month to pay for a year's supply of diesel fuel. Ashamed that you are a month late on Ak Akito's pesticide bill and that somewhere three earlier custodians of your land are frowning? Better yet to be half a year tardy and more: let him, not you, pay the interest.

This is farming nowadays—not a lifestyle, not a noble enterprise, not a stewardship of ancestral ground—and there is no place for the bawl and the whine.

CHAPTER THREE

THE GREAT RAISIN CRASH OF 1983

In retrospect, 1983 started well enough. I knew we were in the midst of the national recession, the beginnings of the Reagan administration's promised, perhaps needed, brake on inflation. But who could have seen not a slowdown but an instantaneous about-face? Who would have figured that in the dramatic shift to deflation and monetarism, there would come also a radical change in the mentality of American business itself that suggested that this downturn was a permanent rather than transitory phenomenon, the creation of a newfound chauvinism in finance, advertising, and insurance at the expense of now pedestrian manufacturing and production? Accountants, bankers, and ad men, the despised of the seventies, old Hesiod's "bribe-swallowing barons" with "crooked judgments," suddenly seemed to hold court, and they all desired a bigger cut of the farming dollar.

True, these coiffured men with their secretaries and pagers could not and would not make or grow things. But they had convinced everyone that, like Mycenaean bureaucrats of a past and failed complexity, their costly legitimizing activities were critical for our present sophisticated and nuanced society and so again they deserved a larger share of the nation's agricultural production. Still, on the plus side, were we not also running up enormous federal deficits, the biggest stimulus package in the history of the republic? Would not all that borrowed money keep people moving, tip the scales to those who produced rather than sat? Agrarian ruin was far from my mind in winter 1983.

More importantly, farmers could take at least one not so good year. A step back for a season, a price freeze, might be good for everybody, might gradually wean some farmers from grandiose notions of expansion on ever more borrowed capital. More comfortingly, the inflation of the 1970s had left the more reasonable with good credit and lots of equity. Our own land, for example, was in excellent shape: equipment adequate, buildings painted and strong, our health excellent, both mental and physical. Also, disastrous rains in 1982 had nearly ruined the raisin crop, cutting supply, drawing down the raisin reserve, and maintaining the all-critical psychological edge in the marketplace, the dictum that raisin growing in the San Joaquin Valley was after all an esoteric art and a tricky business, that the cereal and bread companies better pay and get their hands on short (and shrinking) supplies before the next year's crop was lost as well. At the end of 1982 we bought equipment, built a new raisin shed, remodeled a farmhouse, and were talking about even more new vineyard stakes and wire to come the following year.

True, there exist real laws of supply and demand in farming. I concede that. I once spent a year among blue-chip economists at a think tank, and they proved that to me. But even the most gifted could never answer who, in a philosophical sense, makes that supply and demand? What triggers a rush on particular worthless goods that are not scarce or a disdain for essentials that are not to be found? Who measures that supply and demand? The perception of the public's surfeit and scarcity, its artificial, advertised, and packaged idea of importance and unimportance, also plays a role in the recompense for the farmer.

Once farm prices crash, a climate is psychologically created that for years accepts food bought cheap and sold high. It takes decades to dispense with this idea that the broker, distributor, and food marketer can purchase for nearly nothing what they sell so dear. It requires a war or wild inflation to force them to cough back out a few percentage points of their high-intake profit, to peel away their foul mascara, to reveal that behind the curtain they were tiny, squeaky homunculi all along. The farmer suspects that the Torquemada barking at him on the other side of the phone is in the flesh a stale cream puff, but others will agree only during a rare inflation when a farmer's plums and oranges are now deemed real, the whiny broker's interest and commission helium-laced air.

All the free-market economists I met who lectured on productivity while ignoring obscene commissions, dividends, and salaries, the Ivy League careerists who pontificated about market corrections and the stabilizing, healthy effect of buyouts, shutdowns, and bankruptcies, were themselves quite a sorry bunch. A pampered lot they were, terrified of the ghetto across the freeway, struck dumb by a hammer and nails, left pale and stammering before the formidable blue-collar white repair man. They preached an awfully stern Darwinism. But even those tanned and fit on their nautiluses would be the first to go in any jungle their own models might create. The few farmers left don't care much for liberal or conservative; it's more urban and rural that makes the difference in their eyes.

Despite a few threatening storms it did not rain in 1983. The 1982 losses had driven down the reserve pile, but for the first time in anybody's memory that did not seem to matter. Wineries, given newfound overseas supplies and price-cutting competition, were not buying *any* American's Thompson seedless grapes. Their price in 1981, only two years before, had been two hundred dollars and above per ton fresh, and half the raisin growers had gone wine. Now farmers were all laying down grapes for raisins as their last resort. It was hardly anyone's resort. People wanted raisins even less than they wanted American jug wine. At least, all figured, if you dried your grapes they would not rot on the vine in September when the winery gates closed.

The commercial banks knew of other problems as they pulled in their once-ostentatious horns and exited agriculture. If in 1979 they begged you to borrow "to get ready to feed the world in the eighties" and "to tap the Pacific Rim," now four years later they preached that American agriculture was going the way of steel, and they wanted no part of it. "You've got a good forty days," the polite and embarrassed woman at the local Alma bank told us, "for this note to be cleaned up." The country, however, was not being easily cleaned up; it was suddenly in a descending spiral where people were waiting, not eager to buy, holding out for the price to plummet, not rushing to purchase before the cost rose.

Real interest rates were rising all during the year. We were still paying over 13 percent on our crop loan—over 10 percent above inflation—as plums, peaches, and nectarines, our initial crops of the summer of 1984 (the first year after the raisin crash), were demolished in price, now four dollars a box instead of fourteen to sixteen dollars, as laid-off workers and minimum-wage laborers passed them by in the supermarket. To harvest your crop you had to pay the consumer to eat it. Mohinder Kapoor always went on about the air-traffic control firings and was hopeful because Reagan had brought down inflation to 2 or 3 percent. I pointed out that Kapoor's budget, even with his indentured laborers from the Punjab and black-market and illegal chemicals, was still 3 percent higher than last year—and as we spoke, his tree fruit and raisin prices were falling 70 percent. Would he not, as in the old days, prefer to borrow at 14 percent as prices rose 12? I listened instead to the distasteful old Vaughn Kaldarian, the water stealer, the equipment borrower: "I've seen this stuff before, you better get out of its way." He did. We did not.

Cash—not land, not fruit, not property—was once again king. I had been warned by the elders, years back in the inflationary sixties and seventies, of its scheduled return. In the eyes of Rhys and Muriel Burton, Hans Hendrickson, and my own parents, economics was not a series of natural cycles, not unfathomable financial peaks and valleys, but conscientious decisions by real but anonymous men with capital and power, a species similar to but bigger than those who bought and sold fruit in the East, who had no patience for the erosion of stored monetary capital. Inflation and growth and its accompanying stable of the restless ignorant, like us, who built and grew things, in this paranoid populist view, had for a rare and unguarded moment slipped out of their cages, and now it was time to rein in these muscled and fanged dogs.

After all, when people scrambled and wages grew, when cash circulated and labor beat back capital, the stock market slumped, interest rates rose, and a parade of opulent, spectacled pundits was wheeled out to warn us all of "overheating" and "inflationary trends," to teach us that we were killing each other with jobs and high wages. All during the 1980s and 1990s politicians bragged of upcoming tax cuts, of grandiose schemes to give the average American an extra three to five hundred dollars a year. But in that same period, interest rates stayed high and inflation low: just a one-point drop in the real interest rate by the Federal Reserve would have given the average indebted American twice what any politician promised.

Safeway, I was told as a child, was the "Crooked S." The sophisticated and discerning reader finds the yeoman's simplistic distrust of brokers and merchandisers pathetic, his mind surely paranoid if not unstrung as well. In many ways it is, manifesting itself in knee-jerk disdain for cellular phones, the broker's lemon-yellow pants and pink shirt; hatred on sight for the wraparound sunglasses, pierced ear, and gold chain; and the scarcely contained desire to kick in the predictable breed of car, whether Lexus, Mercedes, or BMW. Puerile though the suspicion be, it derives from a repetitious litany handed down unbroken from great-grandfather to grandfather to father to son. Still, I wager the lore of the unscrupulous farmer is *not* sung in the generational tree of the broker; there are as yet no *Giants in the Earth*, *Growth of the Soil*, or *The Good Earth* written about the majestic struggle of the commodities merchant.

The ode is identical every decade: the collapse of the United Raisin Cooperative in the early 1920s; the sale of two crops for cattle feed after Mr. Habib failed to honor his pledge of sale in the late 1930s; the loss of thousands of plums to Mr. Sledge of Sledge Growers, Packers, and Shippers fame; the clip-and-run tactics of Art Artanian and his fleet of "fruit-wings trucks"; and the sordid debacle of Larry Black and the world-famous Valley Girl Raisin cooperative. Lesser losses (of say, below $10,000) go unrecorded.

Grow up, farmer, get involved, and find a reliable shipper and broker, you remonstrate? There really are none, I fear, after lifetimes of URCs, Habibs, and dozens of others too numerous to chronicle. And remember too that the yeoman's desire for solitude and his pride in independence, the necessary prerequisites for forty years of flapping vine canes and washed-out furrows, are lethal to communitarian enterprises, peer bargaining, and shared food slowdowns. Don't suggest either, as Professor Bart Refino of the university was wont, that the successful yeoman must "verticalize" to capture profit when there is no profit. Are we to believe that twenty thousand men in flannel and denim must jet to and taxi around Toronto and Atlanta, peddling directly to supermarket buyers the produce of their forty-acre patches south of Fresno? We are not, because Mr. Refino apparently means to be rid of those in flannel and denim and their forty-acre patches.

You must be paranoid, agrarian; your farm has survived and your children prospered; these predators with telegraphs and phones *must* have done something right by you. Does not their rising tide raise all boats?

No, no, not all. Prometheus-like, they gave only the bone, with the meat and fat chewed clean. The only thing that has saved farming man is nature. For once in a very great while, calamity saves the farmer through the destruction, the utter annihilation of the harvests of others. About every twenty years there is a horrendous rain, and you perhaps are in the minority who picked early and so saved the raisin crop. Or a killing freeze ruined Georgia's peaches, and most in California were knocked off by hail, yet miraculously yours set heavy and came through unscathed. Even Santa Rosa plums—the poverty-house plums that drove my poor uncle out of farming altogether—can hit the jackpot if your orchard sets when others do not. In these rare, these very rare and outlandish occurrences, for a brief moment, the entire relationship of the ages is stood on its head.

It is astounding to behold, this stuff of a Frank Capra movie, this grower as king for a day, this warted frog as dapper prince! The broker—no, his boss from New York, *divus emptor* himself—calls. They want fruit, want it at any price. To hell with the regular market. The wealthy want fruit, and what little there is in the valley must be on their plates in the hotels and restaurants in two days, come hell or high water. Get it on the truck, they will pay whatever it takes. For a year or two in a farmer's lifetime, the stars line up and the agrarian makes more money than he has and will in two decades combined. That single year, that dream that keeps us all going, is what saves us, never the broker.

I once went over every page of Rhys Burton's diaries, his copious notes from the 1920s to early July 1976, the day before he died. The best I could tell, he never made any real money except from 1943 to 1947. But those five years—the war, the postwar inflation, two crop disasters—all combined to free him of depression debt, to pay for expensive treatment for his crippled daughter, to re-pipeline the entire ranch. Then the spell ended; it was all over, and the Dies Irae returned. Still, that wondrous time had become chiseled within him, become the stuff of legend and local fable ("Old Rhys Burton west of town sent all those girls to college, the crippled one, too"), and so it was more than enough to keep Mr. Rhys Burton going for the next four decades, to make him kind and generous, never bitter and sullen, at the end to send him off at eighty-six, not broke and forlorn as he had so feared in his thirties and forties, but with his land paid for and a little money in the bank.

After ten years of the Greeks, I had forgotten all the warnings about depression and especially the need for parsimony when times, like 1976–1980, were good. Who at thirty has any desire to sock away cash when a vineyard wants stakes or a pump motor needs rewinding? Who wants to lose his legs and collapse just because an ancient reminds you that for everything there is a season? I had never much liked my grandfather's patched bibbed railroad overalls and his resoled work boots. He was somewhat ashamed, after all, even in the last year of his life, thirty-five years after the end of the Great Depression, to buy steak at the local supermarket. Logan, my cousin Rhys, Knut, and I would rather have the profits go to the tractor and the shed to grow more of those raisins than the cash equivalent in the bank to prepare for the fated catastrophe to come.

Another problem was the new grape plantings. During the heyday of the late 1970s, entrepreneurial, but more often silly and foolish, growers wanted to be the big men of their communities and so had financed enormous blocks of new Thompson vines. These are the greedy of every small community, who give inflation a bad name, who wish to apply the ideology of fast-food franchising and leveraged stock purchase to expansionary agriculture. Rugged individuals, would-be Milkenses and Boeskys of the raisin business, planted megavineyards to ride the raisin boom, to reel in prices as raisins "go to 1,500 and 2,000 a ton," as the realtors and vineyard-stake companies forecast. "Bert Kukendahl," the shiny ag magazine ads read, "farms twenty-seven ranches, runs five packing houses, and owns one hundred and fifty-two tractors. He just traded his Cessna in for a helicopter. Bert Kukendahl uses Brand X herbicide. Says he couldn't farm without it. Can you?" To a man they went under, sticking the local land banks and production credit associations, their former darlings, with millions in uncollectible loans. All they left behind were money-losing vineyards worth less than the cost of their stakes and wire, beautiful floral dinosaurs with enormous powers of production caught in a new agricultural ice age.

By 1983 these new bank- and insurance-owned monstrosities were coming into full production—at just the precise time the public was not buying their crops and vast winery acreage was being diverted to raisins. We were not to produce 200,000 tons of raisins but perhaps 400,000 tons and more, just as our domestic market was constricting to barely a third of that number. Demand was back to 1970 but supply was at 2000 levels. Unfortunately it was 1983.

There was more, each week in late 1983, always more—farm disaster is always more instructive than agricultural success. Export sales dipped overseas. "Free traders" in the new administration *welcomed* in European wine and raisins, as if foreign goods freely floating across national boundaries would *force* efficiency in our own native-produced food. In theory, as the automobile industry showed, it made sense; in fact, it was completely fallacious because all our farming competitors were fully subsidized by their governments and their mother trading organizations and so were less efficient than we in the first place. (Only 21 percent of American farms—the majority of them quite large—receive some type of government subsidy.) Selling, not pricing, became the new mantra of the Farm Bureau and Grape League, as if dumping half the crop on the market well below cost helped the farmer.

For the would-be world economist, did it make any sense in the 1980s that rugged rocky hillsides in Greece and Italy, the so-called *eschatiai* of ancient Greek literature, bare for centuries, were rushed into fruit production, while laser-leveled, deep-loam vineyards in California were abandoned or farmed for taxes? Was it right that high-cost European farmers would drive efficient Americans off the land? A once-pompous delegation of conservative raisin growers with "ties to the administration" slunk back from Washington without fanfare to the valley. The rumor went out that some low-level trade official had dismissed them with a curt "who in the hell cares about a few thousand raisin growers?" Who did? Most thought there were such things as raisin plants. A new species of glib government economist surfaced, teaching us all that bankruptcy and oblivion were positive market corrections as world prices bobbed and weaved to rediscover their natural levels of a quarter-century prior.

I attempted my own pathetic lobbying as I watched friends and family slide. I grilled leftist Greeks and even the sly socialist Greek consul general himself in 1988 during a garden reception in Fresno. Would they, I begged impolitely before the annoyed wealthy Greek-American crowd, please expel NATO and all the Americans in southern Europe? Could we Americans not use that savings in their defense to match their own subsidized vintage? Stunned that anyone, much less a raisin dullard, dare expose their con, the Europeans grew hushed. They wanted it three ways: to rail and cavil at an American presence in their land (in deference to the Left), to enjoy free protection from the Russians (in deference to the Right), and (in deference to both) to trade and profit as they liked (for us, the lowly few who grew raisins, that meant to dump fresh, dry, and juiced fruit as the need arose).

Finally, the profits of the 1970s had induced many valley farmers to buy sleek equipment and trellising systems. Heavy, costly manures and toxic nemacides were routinely applied to valley vineyards, along with shiny new metal stakes and all sorts of trellis wires. "I got $2,500 in steel an acre," one fool bragged to the local vineyard magazine. A new generation of very potent pesticides and herbicides was added to the wealthy raisin grower's already lethal arsenal. Every living thing—worms, insects, fungi, viruses, bacteria, in the soil, on the ground, on the vine, in the air—was to be targeted. Forget whether expensive chemical vineyards are sustainable in the long haul. Ignore your own family's drinking water supply pooled not far below in a subterranean lake beneath your tractor. Turn a blind eye to your son on the daily spray rig. Mohinder Kapoor's boy, Eddie, developed an ugly looking growth on his neck. Was his daily fungicide dusting of his trays the cause? Eddie for five years emerged from the vineyard, white with dust from the blower on his back, powder on his clothes, in his truck, powder on his tools for weeks afterwards.

In the short term all that mattered was that these killing fields produced a lot of raisins at $1,300 per ton. By 1983 average grape production per acre—and the cost to achieve that production—was soaring. Always more acreage, always more raisins per acre. Even Fresno's ozone smog and the housing developments could not stop the agribusinessman's chemical counterresponse. As suburban tracts gobbled up vineyard land, as the air of the valley grew foul with an extra million people, raisin production went up as consumption went down. Few sober farmers saw that this was precisely the moment for less (borrowed) capital investment, less production, and more environmentally sound practices. Why go broke wearing out the land for people who don't want what you grow?

The Furies arrived immediately upon harvest in October 1983. A ruined world market, a nonexistent wine alternative, and flat domestic raisin demand, together with spiraling Thompson grape production, all collided in the space of a month. Commercial banks announced by year's end in 1983 that they were leaving agriculture, and all notes were payable immediately. You could finance a pool without home equity, but $500,000 of land couldn't swing an ag extension of six months, so convinced were the banks that the end of family farming was at hand. Any grape ranch purchased for over $10,000 an acre was repossessed within days of harvest by the unfortunate owner or the bank, as cash vanished and the albatross of a money-losing farm was cast back around their necks, as the vineyard slid in the space of a year from the high of fifteen thousand to ten, nine, eight, and five, finally hitting rock bottom at $3,500 an acre. We watched helpless as the tide of 1983 receded, betting each other how low it would go, eyeing nonfarm assets, anything not tied down as an offering to the bank. Everyone looked sidelong at one other to loot some cash, some equity, some capital for the hungry vineyard.

The federal land banks and production credit associations were flooded with new applicants—begging, crying, screaming applicants. It is an ugly thing to watch one of sixty years, with crusted face and hornlike hands, cry. A man who would not whimper as he lost a finger in a grinder will become wet-eyed before a smug twenty-five-year-old finance graduate of Fresno State, flushed with an eerie power dropped inexplicably in his lap. (Were these not like the old despised Scipio Nasica, who ridiculed the rough calluses of the Roman yeoman by scoffing that "he must be in the habit of walking on his hands")? These exalted adolescents turned down almost every farmer in need ("cut out the dead wood"), mourning the numerous bad loans of their own. Lenders now demanded impossible equity, twice and three times in assets the amount borrowed, off-farm income, furniture, houses, cars, anything that was not grown in the ground, anything without leaves and fruit. Real estate agencies went under. Agents returned to teaching or accounting as nobody bought and nobody sold. It was all repossess, mortgage, or rent out. Equipment sales were flat. New tractors sat dusty on empty lots, priced now not at one but five acres worth of vineyard. I comprehended for the first time what Rhys Burton had said about land in the 1930s: you can't rent it; you can't sell it; you can't farm it without losing money.

We no longer just drove tractor, irrigated, and managed but pruned and tied vines all winter and fired most of our farm laborers. In the spring we thinned our own trees, picked and packed our tree fruit, let entire orchards full of fruit rot for want of a break-even price. Four men tried to do the work of twelve. A lot of plums stayed on the tree unpicked and unsold, decaying until the first frost in November. A few acquaintances called us peasants, but we were still alive, and most of them were not. We had other plans for our salvation on our dad's place nearby in Lundburg.

By 1984, at precisely the time the national economy was posed for its artificial boom under enormous stimulating deficits and near-zero inflation, the California raisin industry was all but destroyed. Land worthless, raisins worthless, farmers worthless, a third of all raisin acreage sat idle, the vines denuded by their masters of their fruit-producing canes. Bins of stored and unsold raisins from years past were pledged by other farmers to any who would mothball his vineyard. Under this self-help program many of us didn't pick a grape for nearly three years. Vineyards that bought cars and clothes now sucked monthly wages from jobs in town. Fred Heine, a model conservative farmer, met me on the alleyway one day: "Don't worry, we're getting all the losers out of farming, trimming the fat." It was Fred, not me, who was already alcoholic at the time, Fred who was cut off, trimmed away, swept out from his father- and brother-in-law's operation. Fred's debt might be contagious. Again, when your arm turns gangrenous, you better cut it off before it takes you with it.

As winter 1984 continued, our own losses gradually climbed. Our 1983 raisins were not moving out to retailers. Stacked bins at local yards in January were no smaller than in November. The Raisin Bargaining Association at first said 1983 had been a "terrible" year. Yes, they threatened, raisin prices might dive from $1,400 a ton to the unbelievable 1975 level of about $700 a ton, an incredible, farm-liquidating figure that few farmers really believed.

But Vaughn Kaldarian told me that $700 was a good price. Unfortunately, he smiled, we wouldn't get it. At $700 we were to lose about $100,000 off our 1982 income. By January, raisins went to a mythical $450 a ton. We had now lost another $50,000 by March. All raisins delivered in 1983 were to be paid the 1950s and 1960s price of $450 a ton—over the space of *three years*. Trucks went by daily filled with raisins to the cattle feedlots. Cows, not people, loved raisins. As I write, a decade of depression later, our 1993 raisin income looks about $100,000 less than it was eleven years ago, a million dollars lost in a decade. The year 1993, ten years after the crash, farmers still prayed that their final raisin checks might reach $800 a ton, 65 percent of their 1970s level.

A series of predictable bombshells followed. The raisin cooperative announced that its accumulated capital reserve was now gone. Where? How? Why? The president and the chief financial officer resigned their $300,000 and $200,000 jobs. Expensive vacation homes, generous annuities, sloppy accounting surfaced. Lawsuits followed. We lost another $40,000 of capital retain appropriated by the cooperative; money we had already paid taxes on but had never received and which no longer existed. The bank, which adored the cooperative, laughed when we listed those retains as assets on our financial statements. I am reading these accounts of raisin depression stored in a cardboard box now for the first time in a decade. I never knew my late mother had saved all the clippings: the great market crash, cooperative failure, foreclosures, and so on printed up in the *Fresno Bee* under titles like "Harvest of Failure" and "Rise and Fall of a Farm Family." From scanning these files today and her notes on various plans to squeeze bank payments and farm expenses out of her salary, I know now when her slow-growing but lethal tumor of the head had its genesis.

In our family the response was concerted and immediate. We mortgaged out another $100,000 from the land. My mother and father in their sixties pledged over their entire salary to make up the other lost $50,000; neither thought of selling—had they, no one would have bought. We cut our living expenses. A $720-a-month draw for a sixty-hour week was what I settled on for my family (soon to be five)—and that was too high. After all, when the self-employed capitalist—be he the corner marketer, the sandwich-shop man, the gift-shop owner—does not make a profit, whatever his toil, whatever the hours of his drudgery, his "pay" is simply borrowed money he has not earned. Even the farmer's emotional defense about "honest labor" and "the sweat of his brow" sounds shrill and pathetic, the psychological cargo of failure. He is not at all the revered European agrarian, but the vestigial American agribusinessman who cannot pay his tab.

We cut all the canes off our vines and quit farming grapes, just let the leaves sprout and grow for two years. We did not spray weeds, replace stakes, or buy anything, simply watering to keep five generations of trees and vines alive. Any penny spent would never return. As you will learn, we turned our attention instead to new ventures in table grapes and plums and gave them the last bit of unclaimed equity.

We now paid only the minimum wage to men the caliber of Javier Moya and Arturo Cazarres, who warranted five times that and more. We sold any equipment we could. I was rusty from five years out of school but slinked back to the academic arena, hoping to find cash outside of agriculture so that I could continue to lose it farming, begging to profess for $350 a month, for a single Latin class at the closest college in the valley. *Amo, amas, amat, amamus, amatis, amant. Videtisne in agro agricolam stultum*?

No Latin and Greek to be had? I plotted at starting from scratch a classical Greek program at Fresno State to pay off the debt. You laugh at the absurdity, but what other skills did I have other than growing raisins? Anyway, wasn't planning new Latin and Greek classes like planting a vineyard? Didn't wise old Aristophanes say, "The farmers do the work, no one else?" Didn't Euripides write, "The yeomen alone save the land"? In rambling, pie-in-the-sky memos, I scribbled out the nucleus of a first-rate program to the startled but unreceptive department chairman and dean: classes and tutorials, majors and minors, field trips, library acquisitions, independent studies, extended office hours—the antithesis of everything I had learned about refined and pedantic classics from the wasted years at Stanford, a methodology identical to organizing and laying out a new orchard. In the end, the startled chairman at Fresno gave me one Latin class, contingent on a year of careful classroom monitoring and "the actual certification of an accredited Ph.D."—apparently my appearance suggested the absence of the necessary but silly degree. I dug through my old trunk looking for the proper certification. I was lectured about tracking mud onto the carpets but encouraged to ignore student complaints about unmatched socks and uncombed hair. I was lucky—I could quit borrowing from the bank; I could leave the light ether of the past five years that had destroyed us and enter the rank dark where salvation lay.

My cousin's wife went up to the local IRS for work. My older brother Knut and his wife moved right off the ranch to look for work; they never came—and will not come—back. My twin brother Logan's wife increased her hours at work. My dad, Axel, worked at keeping his huge old Buick alive, to free up his retirement funds for bank payments; my mother began shopping for credit union loans and comparing interest on second mortgages. We sold anything metal on the ranch for scrap—disk blades, tanks, ancestral iron implements, abandoned washers and dryers. So concerned over the growing mentality of scrounging, I was reluctant about leaving the lawn mower—eighty pounds of high-grade aluminum and steel—outside the garage. Still, that outside capital was not enough to staunch the hemorrhaging and save the agrarian life force. A farm that had supported four families would not support one; no, it needed three families to support it.

Is that such an evil, insidious thing that the agrarian should abandon his haunt, put tires to the freeway, and join the other one hundred million Americans at office, factory, and school? In absolute terms, no; for he is born no better, no worse than the youth groomed for the family plumbing business or the aspiring son fueled by dreams of real estate and financial planning. No, it is only evil in the sense of the lost domain, of the red-tailed hawk that perches from the municipal clock, the raccoon that forages in the back alley, or the gopher snake that nests in the pipe behind the restaurant—the squeezed natural order that now appears so ridiculous. When I see the farmer of forty acres behind a desk, I worry not that his belly is empty, his blood pressure untreated. Indeed, his stomach and heart now are no doubt in better hands. The pathos lies instead with the acknowledgment that the last fish is now caught, the last imprint of what we all once were is now smoothed away, and for better or worse, the keyhole view of another world on the other side, the ancient agrarian side, is now plugged from view.

So most raisin growers now more often fail; usually for three equally simple reasons: rain, oversupply, and the traditional bane of all agriculturists, deflation. And as we learned, surplus and deflation are worse than any rain.

Like nuts, cotton, or grain, raisins are a storable commodity. That is very bad now in American agriculture. They can lie in tented and fumigated mountains, thousands of boxes strong, for two to three years. Two back-to-back bumper harvests create not a limited market but no market at all. Who, after all, now eats raisins? Women with gray hair in pinned buns, the elderly few who dislike prunes, the occasional backpacker and nature freak, the rare conscientious homemaker preferring natural to refined sugar in her offspring's mouth.

Go into a supermarket and study the purchasing habits of the average overweight American. It is all subsidized sugar, starch, and fat. The grocery cart is full of manufactured food, bread, cakes, candies, frozen dinners, and breakfast preparations. Most of those ingredients are processed sugar, corn, cereals, soy, and assorted other staples, whose plenitude is underwritten by the government, whose steep profits lie in the processing and merchandising rather than in the growing of the components of those bizarre creations. The true lethal narcotic in this country, the real killer, is not crack cocaine, not tobacco or even alcohol. It is cheap, subsidized, processed sugar and enriched starch that layer with folds of blubber the rear, the belly, and the thighs of the American eater.

In the bewildering months of this 1983 raisin holocaust, I once pathetically followed four enormously oversized people waddling through the Alma Safeway, wagering to myself that their insatiable caloric intake might suggest they were there to buy a box of Raisin Bran, some muffins, or a snack pack of raisins, perhaps my raisins, my family's raisins. Was not the industry tale of depressed sales a sham designed to hoodwink impoverished growers? No such luck. It was Tuna Helper, TV dinners, Pop Ups, Instant Breakfast, Kool Aid, and frozen Sara Lee biscuits and pies they grabbed. Americans have little time to cook, much less to bake. Per capita consumption of raisins in America has been falling steadily since the 1920s. Even the massive increases in population have not balanced out the individual's distaste for raisin bread, muffins, or preserves. Children eat Snickers and Milky Ways, not snack-size boxes of Valley Girl raisins, not even chocolate-covered raisinettes.

Export the surplus, you say? Seek the free market, you overly complacent and unimaginative farmer? Unfortunately, Europe prizes family farming, believing the true price of agrarianism is cheap. In their eyes an ancestral vineyard creates community stability. It inculcates in youth subversive ideas like reverence for the elderly, repairing rather than buying things, physical labor, and staying home in the evening. For some reason they believe an agrarian patchwork in the countryside two millennia old, not a boom-and-bust urban sprawl, lends an aesthetic and sturdy fabric to the nation's character. So the Old World protects its native agriculturists with steep tariffs and generous export subsidies. Greek raisins are supported within and without the European Common Market. Farmers should all leave America for southern Europe for periodic enhancement of their crushed egos. Tell a Greek you're a farmer, a *geôrgos*, and his eyes light up in respect as you enter his inner circle. These are smart men, remember, who can prosper from a half day's work on scraggly vines in stony fields, not, like clever us, losing money from first light to dusk amid verdant acres of plenty. We work harder. We have better land. We are better farmers. We, not they, go broke. I have been waylaid five times in Greece by demonstrating agrarians, their cooperative tractors blocking the highway until more subsidies flow to keep their half-dead pathetic trees and vines alive. More than five times I have heard farmers here smile as yet another crashes or gets caught with too much interest and too little equity. They force a laugh when the neighbor crumbles: "I guess Mr. Bigman, Rocky Giannetta, went belly up over there in Caruthers when he didn't get his fourteen-hundred-dollar a ton raisins."

Worse than European raisin subsidies, cheap jug wines from Spain, France, and Italy flooded the United States in the 1980s, virtually ending overnight our own wineries' interest in Thompson seedless juice. European vintage turned every California Thompson seedless grower (now known as a Thompson "worthless" man) into an orphan, unwanted by winery, bank, or packer. After 1982 they had only the chance of making raisins, and so all laid down grapes for drying, not fresh consumption, not wine, not concentrate. There is a market for 150,000, perhaps even 200,000 acres of raisins; there is none for 300,000 and more when the wine and fresh market force everyone to lay their grapes out to dry.

Is there any need even to mention the trade policy in Japan, Taiwan, and Korea on the importation of American foodstuffs? We sent our Asian pears—Shinseikies, Hosuies, Kikusuies, and Nitakas—there once, beautiful fruit, shiny, crisp fruit. A dissertation could be written on the sequelae of tariffs, phony bug inspections, and transit delays. The Europeans and Japanese don't draw much empathy from American farmers. I grew up with my Swedish grandfather Hans and his wheezing mustard-gassed lungs. They had pulled him off his farm in 1917, and when he was shipped back three years later, he never did much farming again. My namesake was shot down in the caves on Okinawa; my cousin got a bullet above the eyes in Europe (his brother Beldon had his fevered brain burned up in the Philippine jungle), my dad was a near wreck after thirty-four trips in the big fire bombers over Japan, and so on and so on. I would rather remember such things than listen to suited dandies on C-SPAN pontificate on free trade, the new world order, and global harmony as we farmers go broke. You can see why no one in our family has much faith in the fairness of the Europeans and the Japanese, why we know deep down that they are the old Axis who still feel nothing but contempt and hatred toward us, mongrels all. They will never trade food fairly, and I have heard more than one mock-heroic farmer go on that these phoenixes in a half century will require another generation of agrarians to traverse the oceans to beat them down in their own lair.

Far worse than the Europeans or Japanese, the rain, or even the oversupply caused by no rain and steady production is the more insidious virus of deflation. It is the third and ultimate visitation on the raisin farmer, this specter of declining prices and land values that ruined American agriculture in the 1920s, 1930s, and 1980s. My parents and grandparents had said the deflationary depression hit raisin farmers hard in the early 1920s, ten years before the general collapse. Everything, they warned, in overly dramatic, near hysterical language, is bad under deflation. Nothing is good. Hope itself vanishes. And so it did. They were right. We farmers got hit in 1983, again almost a decade before the big job losses and downturns you others in the cities experienced.

Oh, it is true that most farmers now say they "like" such Republican constriction, the hard dollar, low wages, predictable prices, stasis, and all that. I won't argue with farmers that skeptical Republican administrations may be smarter in dealing with drugs, welfare, the lazy and criminal, and other social ills. But raisin farmers, even conservative farmers, usually—predictably—go broke voting Republican, hating the rare Democratic administrations as they become prosperous. So much for *homo economicus*. Jimmy Carter is still a dirty word in the San Joaquin Valley of California, even though the only agrarian wealth here created in the last two decades—new tractors, sheds, stakes, plantings—came under his wild, directionless administration when the suits could not keep up. Raisin prices between 1976 and 1981 were steady at $1,200 to $1,400 a ton. During the Nixon, Ford, Reagan, and Bush tenures the price paid on all the crop never reached $800. It was usually between $400 and $700.

Tell a farmer that: he almost punches you in the face, citing rains, luck, and all sorts of extraneous, superfluous factors for the Carter extravaganza of the late 1970s. He hates you for saying what he knows in his black heart to be true: Democrats inflate and expand;

Republicans deflate and constrict. Democrats enrage farmers with their farrago of entitlement and permissiveness; Republicans excite with their stern talk and get-tough threats. But Democrats make farmers rich; Republicans make them go broke.

Our compatriot in disaster Bus Barzagus returned one weekend from peddling fruit in L.A. after the crash of '83. Reagan was up for reelection the coming year, and apparently a local Republican assemblyman down there named Vineyard was too. Anyhow, someone gave him an enormous campaign billboard of Vineyard's that said REAGAN / VINEYARD in '84. After blackening out the "/" and "in '84" he posted it out in his own weed-covered, nearly abandoned Thompson vineyard. The sign lasted only a few hours before all the incensed, nearly bankrupt neighbors, the black hearts embarrassed at seeing the truth in print, tore it down. Bus, conservative as any farmer around, was now advertising that there was nothing conservative about driving farm and land prices down by 70 percent in a single year. That, after all, was pretty radical.

Are farmers then die-hard conservatives? Most are. Their conservatism still is nonsensical because it is more social and cultural than political, and so it differs from the Wall Street banker and the blue-collar plumber alike. The tight-fisted yeoman spends little because he does not know whether his crops will produce and, if they do produce, whether someone will pay to harvest them. This worrier and doubter does not like change, for he has enough real change for a lifetime each year, every year, as the trees and vines bud out, bear forth, and play out, as crops wither away a week from harvest, as a four-year-old orchard suddenly dies a month before its first crop. The man with feet rooted to the soil distrusts the moneyed and the transient, liberal and reactionary alike, because they are more alike than different, the two having little belly for living in the same house, working at the same job, staying married to the same person, or putting their progeny out to work with their hands. So he votes against his interest, votes solely for the closest thing he can find to a reactionary like himself—this radical reckless gambler who alone of his community will put his entire year's work on the dirt!

Rhys Burton saw the bitter harvests of entitlement in the 1970s, the crime and sloth engendered by state subsidy; in the 1960s he railed against the destruction of the family caused by divorce and promiscuity. Taking drugs, he thought, was no different than putting a gun to your head. And so did he join the chorus of conservatives who were calling for a Republican day of reckoning? No, never. "It's too bad, but I guess I have no choice but to vote Democratic," he said in disgust in 1972. "The other bunch will eventually turn the country over to the fat cats—they always do—and I've already had enough of thirty-dollar-a-ton raisins for one lifetime."

But farmers can handle a Democratic inflation if they are smart and don't go to the trough buying new land, leveraging what they have for superfluous mechanical acquisitions and houses. In this psychological climate of activity, there is a diminution in the value placed on cash. To acquire and save accumulated capital, one must move, be active. Goods and service are what count. Farmers have lots of goods—valuable fruits, vegetables, land, equipment, real things—but little actual cash. How can you when you need enormous amounts of invested capital simply to get through the year?

Under inflation crop prices rise. Land increases in value. Crop loans are paid off a year later in cheaper, easier-to-acquire dollars. People of courage and muscle hustle, scheming, planning to master their trade, to meet the demand for their precious commodities, despising the drones of cash accumulation who will not or cannot fabricate or grow. Strange characters of all races and classes appear out of nowhere to boast that they too are now players in the wonderful American crapshoot; these strong-backed believers will build you sheds, steam-clean your engines, put in new vineyard endposts, relieve you of your extra inflation-spawned dollars. Creatures of a season, they wash in and sprout alive in the high tide of 10 percent inflation.

For the inflation-driven farmer, then, money is just an accounting artifice that the timid and suspicious use to keep score. It is nothing to the man of action, who can produce food and feed people, a great number of people. No one fits into this matrix of hard work, ingenuity, and an endless desire for further production better than farmers, especially raisin farmers in central California with all their weather, water, labor, and talent. No agrarian likes the absence of challenge, the idleness in guarding a pile of dollars when there is additional horsepower, nitrogen, and bodies needed on his land to satisfy the now-insatiable American gut. So former fruit picker, labor contractor, tractor driver Frank Hurtado down the road cajoled, mortgaged, and borrowed to turn his arthritic back and ulcerated stomach into sixty acres of trees and vines, a million dollars of inflation equity that said there was a place at the table now for savvy, tough men like himself, men who liked to pour on the fertilizer, water, and horsepower.

But the entirely opposite mentality pervades a farm during a deflationary climate, which usually predominates in farming. Prices, never expenses, go down yearly. I saw in 1983 the cash-laden and inactive emerge like sated carpenter ants from the woodwork, as the microcosm of the inflation tide pool evaporated. These family and friends, who had been passed over in the shadows, all along resenting the prosperity of ceaseless motion, had been waiting for decades for the moment to preach their gospel, to receive their obeisance; 1983, I suppose, was their moment. "I stay at home still getting 13 percent on my money," eighty-three-year-old uncle Tillford then beamed of the ancient sale of his land, "and I tell you it's worth less to those geniuses now than when I sold it."

They smile as land crashes. They chuckle as weeds sprout, tractors stall, and the so-called big men leave the farm for work in town. Crop loans and mortgages are paid back in scarcer and scarcer cash, as the first volley targets the Frank Hurtados of the valley, who lose their equipment, house, land, life, and soul. Cutting expenses, wages, and supplies is the only recourse for survival, as a meanness spreads over the valley. Tractors and buildings that produce food are fantasized by the suddenly bankrupt back into dollar equivalents. Our neighbor next to our other land in nearby Lundburg, Howard Stallings, had $500,000 worth of trailers, trucks, tractors, sprayers, cultivators, forklifts, and tanks; but when the squeeze came, they were no substitute for $150,000 in cash, and so he too went under. Almost every one of those around us in Lundburg, who owned the best land in the United States, went under.

To the prudent and careful man with stored capital this is all "stability" and "normality," as high unemployment, high interest, and low demand make him rather special after all. But for the trapped and targeted debt-ridden farmer, more than his crop declines in value, as his daring and energy now are revealed as reckless and unpredictable. His very activity takes on the loser's image. His sole cousin in disaster is the miserable farm laborer, the only other player at the bottom of the agricultural quagmire. Booted by the merchandiser, squeezed by the distributor, the hauler, the broker, and the packer, the raisin farmer in turn has only the laborer below to whom to transfer the chain of meanness. For the grape picker and pruner, his own Pavlovian response? Is there not only himself and his family left to beat down?

Deflation inverts the world of the farmer. It says that six generations in the same house are not admirable, nor is belief in an identity free from the modern and material world, but instead proof of a continuous, childish fear of that necessary outside, of an adherence to a failed agrarian regimen that has proven wanting. The choice in one's youth to have passed up better in town for the unfettered life of the yeoman is no longer praiseworthy; no, it is utterly foolhardy, an antiquated idealism that has brought ruin and havoc to everything the farmer holds dear. Once your raisins sell for a third of last year's price, you wonder why, when, and how you have shortchanged your kids and wife, let down the prior planters of trees and vines whose house and land you now inhabit but cannot keep. In 1981 an Alma acquaintance might say, "My God, all four of you boys are raising a storm out there on Rhys Burton's old place;" in 1983 he would snicker over our identical tetrad, "Can't any of you with all that schooling make a living?"

So the raisin farmer relies on his mastery of technique, the excellence of his labor, and the God-given abundance of heat and water. Even the banes of his existence—crop-destroying rain and the oversupply brought on by his very success—still pale in comparison with the entirely human-created blight of depression and deflation, the sudden change of rules in the midst of the game that says no to the farmer, his crop, his land, his labor, says that his mind and body themselves are of no value. The decade to come is to be one not of hopes and dreams but of unending misery as each successive year cancels out in turn the rare good seasons of the past.

But has not this become too abstract, too journalistic an account of 1983? Ruin on the land is never abstract. It's a real destruction of generations of work, a destroyer of mind and body that tears at the cohesion of a family and leaves in its wake the twin sisters of suspicion and distrust. Did the dear departed sacrifice and suffer for it all to end up like this? Do they lie in a single, neat row at the local Alma cemetery for their lesser progeny to lose their life's work as they limp off to apartments in Fresno? Nineteen eighty-three left you wondering how much worse you were than those ghosts in your house who had survived the railroad, the brokers, the wars, and the depressions of the last twelve decades. We would sit each day on the tiny nineteenth-century furniture of Louisa Anna and Willem Burton, uncomfortable walnut and velvet chairs designed for people not much over five foot five, and wonder how these smaller men and women, without electricity and gasoline, could have continued when we had now hit the wall. There were indeed, as we discovered, "giants in the earth in those days." They cast pretty long shadows now, as I read over old letters about a $200 income in 1932 feeding a score of people lodged in the barn and water tower.

Still, for his own protection, the family farmer of today should not have the passionless mind of a historian. He should be no Thucydides or Tacitus matter-of-factly recording the yearly crises in the great history of a state. For if the 1990s agrarian should see his own small history as the ancients might, if he dare compare his deeds with those of his kindred who came before, if he should see his tenure only as a link in the great chain of succession, then surely what started in 1983 conjured up now awful images in his brain. His final chapter might end: "In the one hundred and seventeenth year A.A.C. (*ab agro condito*), in the sixth generation of the line, when the farm was at its greatest size and production, when the raisins had been delivered in the fall, before the first winter frost, as the walnuts were just coming into season and the persimmons were turning a burnt orange, the Burton-Hendrickson clan at last fell into a difficulty that they could no longer surmount. Through poor prices, poor planning, and the decay and decadence of its last generation of wardens, the farm was parceled, rented, and sold and so ceased to exist as before, the work of a century and more now abandoned and forgotten."

CHAPTER FOUR

TO SAVE A FARM

Thus far I have written of Rhys Burton's farm. But we also have owned for a century another parcel ten miles east, my father Axel's place, which his Swedish grandfather Knut had purchased about the same time Rhys Burton was born. It is located in the so-called golden triangle (the realtors' parlance, not mine), a rare belt of fertile farmland between the Kings River and the foothills, the richest farmland in the richest valley in the world. No wonder the corporations and insurance companies call us still to have that forty acres, to pay well above market value for a small gem in their ever-resplendent crowns. And so it makes perfect sense to farmers that we looked to the Lundburg ranch to bail us out of the coming raisin cataclysm on our home place in Alma. But to the other 99 percent of the population, it is sheer insanity to borrow more money to continue in a profession that has nearly devoured you. Still, if we could now begin to farm the better land in Lundburg with crops other than raisins, perhaps that farm could pull out our home place in Alma—after a century the ground of Knut Hendrickson and Willem Burton united by their joint kin into one single operation.

So to save our raisin land somehow, we had gone out on a much bigger limb by taking the best ground in the world, putting ourselves into even more debt, and for seven years (1983 to 1989) producing, as you shall learn, no fruit other than rotten red grapes. That stupid but exalted endeavor, born in careful planning to save our raisins, was now surely to destroy them.

Whom to blame but ourselves? Why tap more of your land's equity to plant even more crops that are not needed? Worse, to lose money in raisin growing involves a certain nobility, with a rich pedigree, an absence of chemicals, and no thought of predation. Table grapes, the biblical "grapes of thorns," are a different story altogether.

It takes a good decade in farming, I think, to learn to do nothing at all. Most often in a crisis, if the nervous, always fidgety agrarian simply freezes, rents out his land, shuts down his operation, refuses to embark on some great notion of production to lead his men on some great odyssey or anabasis, he is better off. Loss of pride is better than losing more of someone else's money. Do factories produce cars when there are no buyers? Do supermarkets continue to sell food at a loss? In an agricultural depression the worst thing the ever-restless, ever-pacing farmer can do is to throw good money after bad, to seek out some novel crusade to produce more food for people who do not want it and who will come to resent, even ridicule and laugh at, its dirt-cheap price. That noble enterprise only accelerates the inevitable downfall, robs three or four years off the land's immune defenses. If the agrarian can strap himself to the porch, recede to the bunker of inactivity, he is far better off than he is marching out in a depression with borrowed capital to tangle once again with the broker, the supplier, and the processor.

But to embrace such stasis and lethargy, as my twin brother frequently reminded me, was to be not a farmer at all. "You're admitting then there's no chance," my cousin Rhys added (I was.). Logan and he both preferred to play Lee at Gettysburg—one last push at the distant center, one last bumper crop of grapes and plums, would yet break us free and clear the march forward. But much earlier, on a single day in winter 1982—the last good year—I counted twenty-one people on the farm engaged in various tasks: pruning, drip irrigation digging, rewiring vineyards, tractor driving, pump repair, mending cracked pipeline, tire fixing and welding—all secure that the checks would not bounce and that we were wise in engaging them to help us produce food. Who and what could feed that horde? Even before raisin prices crashed, their tab was eating at the finite pile of borrowed money at the bank. That foggy morning the bare vines looked especially dormant and not at all up to the task.

In 1983 we could not do what the new laws of agribusiness demanded, and so the already wounded self-destructed without a whimper. Youth like ours brings prerequisite strength to farming, but as Solon, Aeschylus, and Aristotle knew, within its recklessness and rambunctiousness lies catastrophe as well. The Greeks knew it as a race: the frenzied endeavor of the fading older generation to inject and infuse its expertise, tempered with experience and caution, into the young before the very optimism and daring of the new cohort bring them to ruin. So in a depression, I believe now, it is better to be enfeebled and aged; no, better yet, senile, dottering, and immobile.

Arlin Nelson at the local land bank had said as much. He turned down flat the first loan request to plant that Lundburg land. "I hate to see you jump on a sinking boat, you raisin boys. The whole fresh-fruit business is about to blow up," he snarled in his office the last year before his retirement. Mr. Nelson hated inflation in general and in particular those whom inflation had inflated. He had been waiting patiently since the thirties, now postponing even his retirement, for the really big crash which he knew would return. And he was honest in his predilections. "Why not sit at home and keep renting your open land out to some other fool. There's plenty of them. That way at least you won't pay for the privilege of working yourself to death. If I give you the money, the bulldozer will have your vines and trees along with all the others' new plantings in five years." So we hated Nelson for giving us less than half of what we wanted for the new project. Later, I suspect we hated him for giving us the half we did get.

We started out, then, on the new development scheme, the plan to save our home raisin vineyard, not with the necessary $3,500 an acre to get us to production on the new land, but with barely $1,000 an acre—an all-but-doomed, nearly impossible task, I can also see now a decade later, from the very start. The money would have been better spent hunkering down under the raisin collapse. Undercapitalization insidiously ruins expansion plans, as the rural entrepreneur tries to cut expenses that inevitably should not be cut: there was no margin for error should the new vines or trees die or should their novel harvests be scant or unwanted. How can one know whether a bare, foot-long stick in the ground will be alive in two years, will in five become a towering fifteen-foot fruit tree, laden with harvests that do not rot, do not fall, and will be wanted by others who know nothing of the produce of trees and vines?

The other, the missing $2,500 an acre to plant the bare land—and pay for fruit stock, stakes, wire, pipeline, three years of water, fertilizer, weed control, taxes, insurance, labor, diesel fuel, equipment, and chemicals—could be supplied, we thought, through sweat equity, by farming until we could farm no more with the used and the castoff: hand hoeing instead of Round-up herbicide, canal, not pump, water, many diskings with a small tractor for one trip with a big expensive machine, tapping funds from our already shrinking raisin budget. That too was naive, for in America now labor is no match for capital, cutting costs counts for nothing without heavy production and good prices.

Still, in the end we made it close, farmed, I think, as few others could have farmed, for we were young, strong, smart, and without money. But it was not nearly enough. For we were also in the 1980s, and the price of our future grapes and plums, if there had been many future grapes and plums, was to be half and then a third of what it had been a decade earlier, our costs twice and three times as high. As it turned out, we lost big at almost everything we did, lost even more capital and equity than we did with the raisins.

How and why, then, did we ever retain title to the mortgaged land when the adventure failed? Only the unexpected death of my mother, Catherine, and hence the funds from her unknown and undisclosed life insurance policy, her last extended hand from the grave to do what she could not do in the light, for a time staved off complete foreclosure.

But I would rather remember that for a thousand days my twin brother Logan picked me up at six thirty, we drove the ten miles over to the new plantings, and with Rhys, my cousin, worked until dark. Like some synchronized machine, relative clicked with relative as it always should be in farming. For nearly ten thousand hours never a word of rebuke, never the mark of Cain, never an error-induced tantrum, always an unspoken knowledge of blood on blood to finish the job in order to save the farm and ourselves. Like the epic images on the shield of Achilles, we were laborers engaged in group struggle, the finest moments of our lives, a family harmony that even the jaded stopped their trucks to gape at. We were, after all, choosing to be active not passive; in the spirit of American agrarianism, we chose to overcome forces beyond our control through the back, arms, and spirit. Over a decade later, all subsequent labor is mere fluff; in comparison to creating trees and vines, our later toils are mere wasted hours of brag or whine.

Slowly out of a weedy, trashy, ugly rented cotton field, like some New Testament miracle, appeared a beautiful grid of trees and vines, our patchwork of vineyard and orchard created by the sheer energy and cunning of men in their late twenties. How from bare sticks in the nursery can something that resplendent emerge, the only ingredients in its creation brown dirt, water, muscular and mechanical power? How can men without new pickups, with a shot old Italian tractor and an odd Allis-Chalmers in reserve, produce such a canvas?

Which was more arresting, the March orchard bloom of the new plum orchard when millions of white blossoms floated onto the adjoining budding vineyard, its own emerging leaves now chartreuse in the sun, or the summer exuberance when the house was nearly hidden with foliage, the sun obliterated by tree and vine flora? Or was it the bleak winter, when the eye caught the combination of human and natural symmetry, the fifteen thousand trellised stakes now exposed beside the massive denuded vine stumps—lines everywhere, crossways, up and down, diagonally, perfectly straight, offering the stark beauty of the grid or chessboard. We and our equipment might have been ugly, but this vineyard and orchard, our land, from now on was not. It was every bit as good as any corporate showcase. Yeomen, not hired workers, had planted every tree, drilled in every grape stake, obsessively killed every weed. *Haec olim meminisse juvabit*, Virgil wrote, and he was right: in the years subsequent it did bring us happiness to remember these days.

Plots in the same environs with two, three times more money invested in them, absentee corporate owners supplying unlimited capital for their growth, did not match the appearance of our rising farm. For sixty years the ground had lain fallow, as my grandfather Hans struggled with his collapsed lungs, breaking horses and herding cattle in his forty-acre pasture, year in, year out manuring and enriching already rich soil so that its permanent crops, over half a century later, might explode onto the horizon. Knut had farmed vines and trees in the beginning, but his son and grandson, Hans and Axel, as it turned out, had rested the land for decades as if in preparation for our desperate call on its resources.

Given the verdant extravaganza, one chemical company representative stopped me on the road to inquire about our secrets of fertilization. An odd combination of micronutrient? Doses of foliar sprays? Perhaps a secret melange of gypsum, peat, or winter cover crop? I smiled and drove on. How could you just say that the land was between the river and the mountains, that for unknown millennia the river had spilled its cargo of debris onto its flood plain—our farm—that for half a century livestock had deposited their waste over every inch of the ground, that for the last decades a renter had rotated various species of row crops, all in preparation for the eventual establishment of trees and vines? So beneath the detritus of the sloppy farming of the renter, below the weed-infested ground, despite the broken pipes and looted infrastructure, there was the finest soil in the world waiting to be tapped, to produce two and three times the yield of the scraggly vineyards of a century on our home place ten miles north in Alma.

Farming is not just the destruction of the wild; it can on occasion be the creation of something in its place more pleasing, for surely our majestic trees and vines were a sight far better than the aboriginal land itself, for eons the pasture haunt of grass and rabbit. Our vineyard had a look of excellence if not style. If we often questioned the economics of borrowing money in a depression to produce food people did not want, a simple look over the farm vanquished all such doubts. Indeed, after ten hours of pulling Johnson grass or stringing vineyard wire, we were loathe to go home. The visual power of its creation held us firm, made us alive, told us that men who could fashion such living things, bring them into the world, could do almost anything and could never be altogether evil. We even had a confidence in our walk, we who had planted a vineyard and an orchard, enterprises that many tree and vine farmers their entire life do not attempt. Each of these rare days I cursed the prior three thousand spent in university towns and strange countries, nose in some bizarre text, brain full of column drums, irregular verbs, and pottery shards; those thousands of hours away from the shovel and pruning shears I knew now to be wasted and lost for good.

So those young Royal seedless grapes and Regal Red plums—the most promising of new species, on the cutting edge of agricultural breeding and selection—made us see that agriculture is not the tired litany of chemicals and compaction, of pollution and desecration. They lie who reduce farming to mere exploitation and exaction. Farming is also the artist's brush stroke, which places an orchard there, a vineyard over here, an alleyway separating the two slightly off center on the canvas, which has the red sun on the Coast Range silhouetting the trees or daybreak over the Sierra Nevada glistening off the vine's stakes and wire. Who needed to make money, to pay back loans, when they had been given the gift of spawning such bounty, of growing taller, greener, and healthier trees and vines than any around? Of course, then we knew not that "we had planteth a vineyard, but we would eateth not its fruit."

The eastern San Joaquin Valley can grow a lot more than raisins. It produces any food really, but especially crops that require heat and water and cold winters. That means just about all species of fruit tree or vine. Theoretically, those wide parameters of the possible allow the valley farmer to diversify. If his raisins are bad, his peaches or plums carry him. If that volatile tree-fruit market collapses, steadier raisins tide him through the year. Theoretically, I said. In actuality, after the agricultural depression of 1983, *everything* was bad, from plums to nectarines, from walnuts to persimmons, from canning peaches to kiwis and almonds. The degree of a crop's lethality depended largely on how much borrowed capital each required to reach harvest. On that chart of doom, tree fruit and table grapes unfortunately headed the list. Any who farmed those high-stakes, capital-hungry, labor-intensive crops soon learned that 1983 was no recession, no downturn, but a genuine agricultural depression of the 1930s type.

Odd it was that we never saw the D word in print, never heard "depression" even spoken among agricultural economists or farmers themselves. There were a lot of "downturns," "recessions," "adjustments," and "corrections," but even when prices crashed 70 percent, to below their levels of twenty years prior, no "depression" was uttered. By 1985 when I entered the local grape supply yard, its robust business completely gone, a single man was left selling, not vine stakes, but chemically treated wood for the suburbanite's deck. The beaten-down owner, Ed Beloian, once boasted of expansion into metal stakes a mere three years earlier; now he was left alone behind the counter. Still, he had enough fire left to bristle at me when I snarled, "It's been quite a *depression*, hasn't it, Ed?" So, too, when Ak Akito (before his pesticide customers began to stiff him) beamed about the dollar's new recouped worth after a decade of inflation erosion, I could only bring him down with the D word: "That's what usually happens in a *depression*, Ak; cash is king."

We four had decided, remember, to grasp at something else besides raisins to keep us alive. For years we had eyed this rented-out forty acres of highly productive land southeast of Lundburg, and this was the siren that was now to draw us onto the rocks. By 1983 we were about to harvest from it our first crop of red plums and red table grapes, just in time, or so it seemed, to bail us out from the raisin fiasco.

Place, as I have, some Regal Red plums or Royal seedless grapes on a city dweller's table; his eyes bulge, so stunning is their hue, their impressive girth, their natural fragrance. He wants them now, as many as he can get. But producing delicious fruit and producing it at a profit are usually mutually exclusive propositions. The rub, as I have said, is that you have to make money in farming; it is no longer a way of life.

It takes a lot of money, borrowed money, to plant a vineyard or an orchard, to farm it for four to five years until you get your first good crop, to win back from the ground enough cash above your farming costs in those first years of production to pay back your initial capital investment and its interest. Vines and trees, as the banks and university professors nonsensically both warned and advocated, are a gamble far more risky than anything in the casino, with a potential for humiliation and failure unheard of at the track. At least it can be quick there. You can wear nice clothes and see pretty women as you lose your roll, as the guillotine falls. You do not pay hourly throughout the year for the monotony of driving the spray rig as you flop endlessly like a fish on the quay. I had never understood agricultural depression and yearly farm losses in quite those bleak terms until Olin Carlson, who owned the local brake shop in Alma and also farmed a small ranch, asked me: "Did you like farming for nothing this year, paying the bank to work your own place?" He lumbered off, laughing to himself before I could come up with a smart reply, his back bent double from years under his rack. He knew he was under the cars every day now to pay for his tractor ride in the evening. A decade later a housing tract saved him when the local junior high school was built across from his soon-to-be-obliterated farm.

If people are born into viticulture and arboriculture, they never stop to consider what an odd, absurd proposition tree and vine farming is. Such farmers see the stern silhouettes of their ancestors, remember the no-nonsense talk of a Rhys Burton, and never make the connection that beneath that conservative veneer of these farmers, they must have the blood of gamblers in their veins. You work from dawn to dusk; you drive a straight furrow and don't skimp on your ditch water; and therefore you think you are responsible, even reactionary, in your allegiance to past protocol, timid in your reluctance to vary from the mean. But any unbiased outside observer who analyzes your enterprise would never join your efforts, never invest in your labor, never lend for your cause. He knows that putting sticks in the ground and expecting to make money from them five years later is a fool's delusion; he knows that allowing two hundred tons of raisins to lie unprotected in the dirt is the stuff of the Las Vegas high roller. A Fresno judge once told me, "Your mother Catherine abhors gambling, thinks it should remain outlawed, doesn't even approve of slot machines, office pools, or card rooms, lectures me about betting on the Super Bowl. And yet here she goes every year and pledges her entire salary as collateral on the hunch that it won't rain or hail, that some broker in New York is going to pay what he should."

The problem was that after 1983 you could not even meet your expenses on a mature, paid-for vineyard or orchard. That was not because the public did not want your harvests or the fecund valley was better off as housing tracts. Quite the contrary. The San Joaquin Valley produced prodigious amounts of beautiful fruit that the consumer welcomed, no, took for granted, at a cheap price. Bounty, surfeit, overproduction was the ruin: the consumer wanted his grapes and nectarines, but he wanted and got them cheap. Unlike car assembly lines or even cotton and corn land, you can hardly shut down or idle the orchard. It won't go into mothballs, unless you're a corporation on the west side of the valley that closes up and sells its subsidized water, your water, to thirsty L.A. for millions in profit. So even if the small guy doesn't harvest his fruits, he must pay to keep the trees and vines alive. And more often farmers kept harvesting their surpluses for want of a better idea, for the reluctance to bulldoze and ignite magnificent producers of unwanted fruit. In their defense, it is still a foul sight to see kingly trees, hundreds, thousands of them, turned topsy-turvy, their roots skyward, enormous holes in the earth, leaves brown, rabbits and squirrels immediately moving into their soon-to-be-torched remains.

If a producing vineyard or orchard was a guaranteed money loser, much less, then, was it feasible to pay back the money borrowed to plant such trees and vines in the first place. That axiom—unquestioned, undeniable, unchangeable, irrevocable, as Arlin Nelson warned—in itself meant failure. But blindly we pressed on and so entered the nether world of deficit spending and steady "capital erosion," tapping, not saving, the land's residual equity built up over sixty years from the last depression. Each day for three years money went out without a crop to bring some back in. And in our case, these particular new species of trees and vines—our salvation, Regal Red plums and Royal seedless grapes—were not what was promised and so eventually involved the ruin of absent production in addition to the squeeze of low prices and high costs.

These fruits were a particular type of Charybdis, a whirlpool that sucked everything—health, capital, tranquillity—in their wake, that threatened to destroy not merely their creators and themselves but everyone, everything that their creators possessed and held dear. Beneath their beauty and extraordinary taste, Regal Red plums and Royal seedless grapes were satanic creations which devoured all the equity fed them.

I prayed that every uncle, aunt, cousin, and close friend stayed clear, kept their checkbooks at home, and their bank statements private; I prayed all that even as I hoped they would step forward with a check and say, "Keep on, you're going to make it yet." You must watch your own in farming now! Out of desperation, seeking either sympathy or admiration for his own melancholy agrarian enterprises, the family farmer will ask for loans and assistance when he should not. He will drink dry anyone in the extended family who would sign the loan guarantee, tear off their social security, pension, savings bonds, car, boat, or house. The farmer reasons that they should lend succor to their own blood when he produces food and they do not. He muses that the money of others, earned outside of agrarianism, should now save the last generation of family farmers. Finally, we became so mad and unbalanced at the depths of the catastrophe, we thought Uncle Tillford himself might let loose with one of his many

$100,000 CDs—as if a tightwad who had grown opulent on the misery of others might lend anything, as if we had any right to accept it if he had.

I am glad, as I write in 1993, that the magnificent Royal seedless grape vines have now been mostly yanked out by farmers, exterminated root and branch by the men whom they sought to destroy, soon to vanish as a species altogether from the face of the earth. Environmentalists must recognize that some of life's species, like smallpox, particularly those hybridized by man, ought to perish, have no business, no purpose, on this earth. Each time I drive now by a rare Royal seedless vineyard, marked inevitably with a pathetic For Sale sign, I still dream of megatonnage, of enough ordnance to send that grid of vile plant work into oblivion. (In defense of its university parents, later rumor had it that the Royal seedless grape had failed its years of testing, had its file closed with the verdict "chemically dependent" and was about to be aborted, its existing young stock burned. Only stealth and the connivance of food brokers pulled out the stake and brought the grape into the daylight world of commercial production.)

Even if the price had been good for Royal grapes and Regal Red plums, even if the demand had been as it was in the 1970s, these two varieties, these choice new species of plum and grape, still would have broken, and in fact did break, any farmer foolish enough to plant them. Born out of the laboratory, the strange concoctions of plant alchemists, neither species could be grown under the real conditions of the vineyard or orchard. They were fruits whose dazzling harvests required either a secret and unknown process of pollination or a permanent fog of toxic chemicals.

It is hard to plant a vineyard in the valley. Clear the land of its last crop, and hope it is free of perennial weeds. Then mark it off into rows and files, one enormous checkerboard, twelve-foot rows all across, six-foot spaces between vines down the row. Make an error in your spacing, and later on when you drive down the row with your disk, the row mysteriously gets narrower, twelve feet to eleven feet six, as your blade has no where to go but to slice up vines. On the next defective row the space logically grows, so by row's end it is twelve feet six, and the disk for the rest of your life misses a good half foot of weeds.

Stick in rooted vinelets every six feet a row, about six hundred vines per acre, six thousand every ten acres. That is the easy part. Within hours they need water, and that means weeds. Without any shade or trellis support the infant vine grows along the ground the first year, mixing it up with the unwanted flora that also take off in the sun-exposed field. You must go through about every two weeks after each irrigation and weed around each rooting—six hundred of them every acre—careful to cut down only the noxious plants, watchful not to decapitate the young, near-invisible vine trunk. Irrigation and weeding are manageable and quick if your field is level and free of shrublike Johnson and orchard grass. Ours wasn't.

Our renters, one now dead, the other bankrupt, had milked the place for years and left it a jungle, an encyclopedia of weed species, varieties whose names I did not even know, whose presence in the valley I have not seen since. When we turned the water into the vineyard, it did not run down the row but pooled and collected like one enormous rice paddy, taking hours of shovel work, wasting water, and spreading weeds. Gargantuan Burt Connolly, with a fleet of dozers and carryalls, badgered us to laser-level the entire place in two weeks. But who had the $10,000 to run his machines, and rumor had it he was about to go belly up anyway? (Nine years later, when he resurfaced after bankruptcy and there was some life insurance money left from my deceased mother's policy, he did flatten the entire place, sent his big machine over the Regal and Royal crematoria, smoothing the land flat like it should have been before the orchard and vineyard were planted.)

We were on our knees, weeding and wading into the water the whole first summer with shovels and hoes, sharpening them with the file every four or five hours. The renters had also plugged the pipeline, scraped the slope the wrong way ruining the fall, burned out the pump, and in general left us with an agricultural ghetto. They finally haggled over the rent, arguing perversely that under their tutelage the place was getting old and unproductive and thus justified special reduced terms. With a limited line of credit and unforeseen problems, reclaiming that weed-infested and unlevel land for vineyard and orchard was a contest between vanishing capital and daily catastrophes. The worse our obstacles became and the more our money vanished, the more the vineyard rose all the same. There were moments when the sheer power of our arms and backs suggested that we might create a productive table-grape vineyard after all. The land, despite its poor custodians of the past two decades, was rich, richer than any field around. And is not the land and its soil, as the Greeks knew, always the ultimate arbiter of agricultural success?

Why plant Royal seedless grapes in the first place? Most every other reckless and desperate raisin farmer with any money left was planting Flame seedless, the crunchy July grape that ripened early and was selling well in the stores. That, I confess, is a sensitive story in itself, better to be passed over quickly. My brother's brother-in-law worked for Effin Brothers, run by an ex–food broker down from San Francisco, Hugh Effin. (He had no brothers; that was simply yet another marketing device to counterfeit a long family tradition and cohesion in agriculture.) Hugh Effin had cornered the Royal seedless market and boasted—I heard this secondhand four or five times—that this esoteric, hard-to-grow grape would never be overplanted. A recession-proof grape, Effin believed Royal seedless was. He said (or rather his field men told us, "Hugh says") that it took the clever and courageous, not the timid and weak, to rise to its peculiar (and still mostly unknown) challenges, to win that privileged place of profiting in a depression, of bringing a rare culinary delight to an appreciative, selective, and wealthy consumer. He had even formed a Royal seedless growers "support group," a captive audience of victims for his agricultural theories, who had promised to turn their crops over to Mr. Effin to be pooled and sold.

Nothing is more dangerous in farming than the philosophical fruit broker. A veneer of intellectualism and wit disguises the venom of his bite until the marketing toxin has done its work. He despises as brutish or unsophisticated the only good thing left in farming, the earth-bred honesty and lack of artifice of its practitioners. I went once to an Effin grower group session but left early after watching passive farmers nod as Hugh conned on about the European theory of grape trellising. Anybody who name-dropped university agricultural professors—Honorary Doctor of Letters Bart Refino especially—and mentioned the landscape of southern France as proof of his pedigree, as proof of his trustworthiness, could not be trusted.

The Royal seedless was a late crimson grape that ripened in September and could last in cold storage for months. You picked late in September (rain or no rain), put it in the cooler, and then supplied the United States with "fresh" fruit during their Thanksgiving and Christmas holidays. Everything else from California was but a long memory from summer past. After eating cheap grapes all summer, Americans would pay a little more to have their crisp scarlet berries continued in the fall and winter. We acknowledged that far more people ate fresh grapes than dried raisins. We nodded when Hugh Effin's team talked of the "Thanksgiving table" and the American housewife's desire "for some good-looking fruit at Christmas."

Bank skepticism or not, Royal seedless, I thought, was arguably a sound proposition, especially when we had just enough equity left in the land to get it planted and going, especially when there was a family tie of sorts with the Effin people who were marketing the grape, most certainly when we could count on no raisin income for the next decade. Many growers were apparently scared of Royal seedless, so the demand/supply picture that was ruining agrarians in every other crop did not necessarily apply to this late red grape—or so at first we believed. Before 1983, Royal seedless grapes were selling each year for twelve to fourteen dollars per twenty-three-pound box. Good growers were harvesting, they claimed, seven hundred to one thousand boxes an acre, and even their rich lands were no match for Knut and Hans Hendrickson's ground. So you can see the type of gross income that was going through our raisin-ruined minds: over $10,000 an acre, $100,000 every ten-acre block. In 1983 that was now ten times and more the gross proceeds of a raisin vineyard.

As soon as we planted, various Effin people visited every week to inspect our technique, to patronize us hard working but capital-poor wannabes. Our fields were clean, vines thriving, stakes rising. These Effin farming capos mulled around the vineyard, outfitted with green hats with a red grape bunch on each hat, grape-leaf belt buckles, monogrammed E pens, pen holders, calendars—the whole brave new world of California corporate farming. The green hats themselves were mostly either failed farmers or failed brokers, the parasites who liked to work off and around authentic farmers without either the toil or risks. Like moths around a flame that will shortly incinerate them, they were fascinated by Hugh Effin, incessantly quoting him chapter and verse, particularly his southern French vacations and anything that smacked of a sophistication usually reserved for the more elegant boutique scum up north in the Napa Valley. The field men wore the refined regalia of agriculture—boots, hats, elegant western shirts; we wore sneakers, baseball caps, and T-shirts.

All we had to do was spend the money, farm the vines, and three years later they would store and market the grape and so give us those lucrative profits. "We're not so interested right now in making profits," Bob Forrest, the chief Effin ramrod lectured, "as capturing market share. We need your grapes as much as you need us, and if we can't be fair with farmers like you, we won't have enough product to get the message out about our own Royals." That sounded good, and besides, just in case of any funny business, our family connection would have an eye out for his own blood, a mole inside the walls with a nose for box switching, label doctoring, invoice rigging, record altering, pallet disappearing—all the orthodox stuff that supposedly goes on once the farmer's produce enters the shadowy world behind the icy barricade of the broker's cold-storage plant.

Remember, in the current mad world of American farming, the company brokerage men wade unannounced all year long through your field, arrogantly race down your alleyways, park and chat at will in front of the house of their serfs. But once your fruit enters their compounds, you cannot set foot on the premises. Suddenly the notion of private property arises. Oh, you still own the fruit—that is, the risks that the stored fruit may rot or not sell for its six-month life in a box—until the grocery store receives it. You cannot visit your grapes unless accompanied by a warden and guards. How foolish to believe that our in-law worked for us, not them! How silly to believe we were peers and not helots!

What, pray, does the *emporus regalis*, the salesman, and the distributor do? It is rather easy to describe and also need not detain us long. The farmer alone grows the fruit. He picks it. He takes it into his shed. He sizes it. Then he puts the graded produce into pretty cardboard boxes with catchy names like Sun Diamond or Kings' Harvest. Next, boxes are stacked ten high on his wooden pallets and then strapped tight with his metal bands. The agrarian forks his year's work onto his truck and unloads it at the broker's nearby cold-storage plant. At this point he bows out, and the others emerge to begin the profitable work of bringing the food to your table. A sixteen-wheel refrigerated truck picks up the fruit and hauls it back east to a produce terminal. It is then resold and trucked to a grocery store, where the colorful box of the farmer is opened and dumped onto the produce section. The fruit, remember, has not left its box, nor the boxes their pallet; *no one* has packed, processed, adapted, modified, even touched the farmer's creation. But in the process the local valley broker who guided that transaction by phone takes 10 to 20 percent of the wholesale price. The trucker, too, takes between 10 and 30 percent, depending on the value of the fruit. The distant distributor and grocery store divide the rest.

That rest is considerable, so considerable that grocery stores and produce distributors find the marketing of fruit their most lucrative endeavor. After all, it is an enterprise devoid of almost every risk. A box of plums that costs the farmer seven dollars to grow, pick, pack, and box may sell gross for eight or nine dollars, in a good year leaving the farmer with a one- or two-dollar profit to buy equipment or pay his nonproduction expenses like insurance and interest. (The years of the 1970s and early 1980s where tree fruit sold for eighteen to twenty dollars a box are over.) The broker and trucker on the West Coast each get their dollar or two, and then the real markup goes to those in the East, where the nine-dollar, twenty-eight-pound box suddenly becomes precious metal. Plums sell between $1.80 and $2.50 a pound, or between $50 and $70 *a box* retail to the consumer. Somebody has taken fifty to sixty dollars per lug, the lug that is merely placed on the shelf and cut open.

But, the men on phones complain, are there not hazard and chance? Just what is the risk, the farmer counters? Is it that their truck might get a flat, that the broker's cellular phone might go dead? Might a power shortage in Chicago cause rot at Safeway? Or is it that worms will eat the fruit on the tree, frost will destroy it in bloom, rain will rot it before harvest, men cannot be found to pick it? Or will the orchard and its harvest be improperly pruned, watered, fertilized, and thinned somewhere between Fresno and New York?

And under the terms of sale, the farmer *owns the fruit until it reaches the store.* All along the chain, any misstep, any catastrophe, and *the farmer alone pays.* We have had the price "renegotiated" on plums in refrigerated ships going to Korea. Peaches have mysteriously "rotted" in the four-hour chilled ride to San Francisco. Checks have bounced and charges have been added for unsightly boxes, bruised fruit, and pilfered nectarines in transit, all of which were firm fruit unscathed when put on the truck. The ultimate horror is the fruit that does not sell and thus is sent eastward into the black hole of "consignment," where the buyers pay whatever pity dictates. Usually in the 1980s the price of charity was a dollar or two a box.

Even for the skilled broker of the valley, the lesser counterpart of the East Coast distribution companies, for the man of the caliber of a Ross Lee Ford, the profits can be considerable, the take high. Should he sell two million boxes a year, two hundred thousand of those become his own commission, free and clear. Ross Lee Ford's office and freezer become the instant equivalent of a three-hundred acre tree and vine farm, void of all the risks and without the farming expenses. His phone grows two hundred thousands boxes without the frost, the worms, the grease, and dirt, two hundred thousand that are always sold, always delivered, always fresh; for on the air-conditioned, adobe-veneer farm of the broker, it is always temperate, well watered, and fertile. I made that point once to a brokerage acquaintance; he laughed when I went further and said those two hundred thousand boxes might make a Ross Lee Ford between one and two million clear a year. "Do you think we work at it so hard, just to make that little bit?"

Do all strive to become as Mr. Ross Lee? Many do. Most fail. The life of screeching on the phone, haggling over price, staying one step ahead of the rival suit, lying to the farmer, having the distributor lie to you, is not an enviable one, and it takes patience and tact and savvy. Once Ross told us how worried he was about our flagging spirits, concerned that we had become so dispirited over tree and vine farming. "Don't worry, I'll take care of you one way or another, whatever it takes." And on the next sales invoice my cousin spotted a $7,000 credit—Ross had decided to share the loss, after all, on a truckload of our plums he had wanted but then dumped in Canada! Ross Lee Ford did care that his edifice and plant were rising as his farmers were sinking! For a week there was a solemn acknowledgment that Ross finally had seen the light; he had heard of our plight and must have known that the measly $7,000 made all the difference in the world to our bank, meant to them that we were finally zeroing out, not losing once again. But then the next invoice arrived and our child's fantasy evaporated. The $7,000 was a clerical error; not only was the money now re-deducted from our final fruit sales, but Mr. Ford had decided that the necessary fruit inspections he had demanded from his plum growers could no longer be subsidized by his generous purse. About $5,000 in "fees" without warning or notice were added on to the invoice, and so we went from being up $7,000 to down $12,000 in an afternoon, finishing the year in "an eroding position" indicative of "continual inability to establish a pattern of profitability."

But it is not gentle, this Ross Lee Ford galaxy of take-out food, cigarettes, and high blood pressure, with the flotsam and jetsam of broken marriages, unused vacation homes, and jet skis for ungrateful children. It is no picnic, the life of the successful broker. How else can I explain why the tottering farmer, on the precipice of bankruptcy, mutters only a little when his fruit is stolen before his eyes. So lucky is he to be away from acres of refrigerators, secretaries, ties, and Lexuses that he wishes only a small return if it may mean only that he can stay so far away. To my carping and whine over pages of deductions and consignment sales, my brother ends the conversation with only, "Would you like to be a Ross Lee Ford?"

Still I dream: in a perfect world, in a world beyond this one, will not the fruit broker one day share the farmer's risk? Once he solicits the grower's fruit, he shall either pay cash right then and there or take his commission on the eventual sale; he shall share fifty-fifty all boxes that spoil or rot unsold in his warehouse and on his trucks. He shall be stuffed and sated with a 4 percent commission on his sale, not still hunger for yet more after 12. His pressed suits and polished nails shall take no subsidy from plums that have been picked by better men of filth and vermin. In that world to come, when the price collapses and the fruit does not move, he shall not send the farmer a bill for services rendered because there will have been no services rendered. When he visits the agrarian, he shall knock on the farm door, he shall call in advance, he shall never race through fields that are not his, just as the raisin grower, the plum farmer, and the grape man now never walk unannounced through his cold storage or browse through his monstrous creaking raisin machines. The broker of harvests shall pay some homage to the man who gives him life, the food producer who entrusts his year's work into his own softer hands. Never more shall the broker and distributor ask that those greater than they, their real lords and masters, kneel and bow.

And finally, do the Ross Lee Fords of the world ever lose money? They do, for they sometimes buy farmland. Glutted with cash, their accountants and attorneys nervous about taxes, zealous for the aristocratic badge of "landowner," the broker sometimes crosses the green line to gobble up orchards and vineyards from the squeezed and exhausted farmers. For the first time in their lives, the bottom line is ignored, and the Ross Lee Fords choose to produce what they can steal from others. But wondrous to behold, no longer can the broker and trader loot entirely from the producer: He is the producer as well now. So he grows food at an unending loss, incessantly plundered by his ilk. He audits the books on his money-losing plums and peaches; he complains about the costs of chemicals, labor, insurance, about the cut from people like Ross Lee Ford; he seeks out depreciation and write-off but finds in their place hemorrhage and havoc as well. And if the ego be too oversized, the land grow too large, the farm output too costly, the broker cannot transfuse from others fast enough to replace his own bleeding, and so he too crumbles, loses his mahogany, his Lexus, his three-acre refrigerator all for the upkeep of a few thousand trees and vines. As he whines over the red ink on his nine hundred acres of nectarines and plums, particularly the dismal prices, I suggest to Ross Lee Ford that he find a new Ross Lee Ford. I suggest to Ross Lee Ford he better get bigger or get out.

Bus Barzagus, our neighbor in Alma, only shook his head at our predicament with the Effin conglomerate. He drove over (glad to be free from a day on his new mountain orchard), took one look at the entire Effin operation, one look at this green-colored bunch of Royal field men in our field, my brother's brother-in-law included, and concluded, "I don't care if your brother-in-law is tied up with these guys, get out now. Cut the vines down, graft them over to Thompsons, and you'll only lose money, not your ranch, farming raisins. I know this Effin and his whole Royal operation. It breaks people. Royals always rot on the vine. It's too costly to get the good bunch. You pay for the rot, they take the few grapes that make it. Good God, they even have the right to tell you when you can pick your own fruit. They even own it when it's still on the vine."

He went on and on, rapidly but expertly reviewing the inherent problems of cultivation, the impossible economics of production, the unfairness of the marketing, the expertise of Effin, within minutes shrewdly extrapolating, as was his custom, a sophisticated, complex thesis whose argumentation no one could contest, much less refute, leaving us his audience at once bedazzled, ashamed, and depressed. He then stormed off, growling at the Effin brigade, depressed and angry that we too were now hopelessly caught in the same net that was smothering him. Circling out of sight, he would burst again into our midst, kick the dirt, blurt out yet another unperceived con, and then storm away. Somehow the gallant crusade to save our ancestral raisin homeland by planting this absurd species of table grapes had been reduced to the gullible's weakness for the carnival barker.

But again, can I for a bit longer return a few years to the planting of the new ground? Even now, over a decade and more later, I wish only to think of the heroic efforts at establishing the vineyard, to dwell once more on our labor and harmony that made its splendor possible. During the first year the vines went into the ground (after all, there were no grapes yet), everything went unusually well and suggested that labor, after all, could substitute for capital. The deep loam of my Swedish great-grandfather Knut's place, as I have said, grew Royal vines unmatched by Effin's most prized estates (in our defense, we were not ignorant in the arts of cultivation, irrigation, and fertilization). The yearling vines were over two inches thick at the trunk, and the canes spread ten and fifteen feet along the ground. No soil in the country could match that place. That winter for three months the three of us hammered out our own vine trellises; nearly twenty thousand seven-foot stakes, with three-foot cross-arms nailed to each, all driven into the ground with our wondrous own homemade hydraulic cannon. What a spectacle that was, mud splattering everywhere as stakes were drilled in succession. Each week more and more rows of 140 wood crosses went in; sixty miles of wire were strung through the telephone pole–like trellises. We worried that one of the 140 trellises might be too high or low, might veer an inch out of line to the right or left, and so ruin the symmetry and strength of the stakes and wire.

Art Medford, the local carrot and cabbage king, watched our work each afternoon from his pickup, taking notes on our rate of fire, curious to see how many stakes we could drill in an hour, marking the cost carefully for the vineyard that he was about to plant and that would nearly break him six years later. Every ten days the young second-year vines were tied up the stake and then onto the wire; we walked daily through the vineyard, making sure a tender shoot or tendril had not broken off the stake, careful to add an extra tie where the vine was not quite snug against the wood.

By 1983 the vineyard was finished, ready to produce thousands of three-pound bunches of Royal seedless grapes for hungry Americans, ready to impress the Effin team with our expertise and diligence, ready to save our land and ourselves.

Vineyard maturity came just in time! We had arrived at the end of our $1,000-an-acre credit, and our raisins had nearly broken us. But now we had a towering vineyard (and a powerful new orchard beside it), whose inaugural harvests were but months away. These new fruits were the cavalry on the horizon to save our farm. Arlin Nelson had spoken too soon!

CHAPTER FIVE

BUS BARZAGUS AND HIS MOUNTAIN

Before disclosing the denouement of our new development venture in Lundburg, be aware that we were not alone in our desperation. There were other valley farmers, hundreds more, who sought some salvation when raisins dived from $1300 to $450 a ton and tree fruit crashed a few years later—who, in other words, had no desire to auction off and go into town. For most, the odds were worse than in our own plight: There was no off-farm income to be tapped, little equity in the land to be siphoned, and no additional and superior plot of ground to be planted. And so quite a few went broke and have disappeared from the rural landscape.

Other agrarians were so determined to stay in farming at any price that they took on desperate enterprises that lost what little equity they had left; they were quite willing to violate all the canons of California agriprudence, the century-long corpus of folk wisdom that tells us farmers what we can and cannot do with the land. Those laws, like the agrarian legislation in Plato's *Laws*, in their aggregate are meant to be inviolable and are privately acknowledged as such. Any farmer who, like Bus Barzagus, would test his mettle, however noble and desperate his cause, suggests that he knows the nature of the land, its people, the system of agriculture in America, nature itself, better than the millions of California farmers who for a century learned through ordeal and catastrophe and passed their hard-won acumen on so that others might avoid their tragedies. In short, Bus Barzagus planted trees on a mountain in violation of the agricultural laws he had honored so carefully on his home raisin ranch in Alma, the very laws that once made California agrarians the most successful and productive farmers in the world. Rather than tell what the mountain was, it is better first to describe the world of agriculture that it was not. Twenty-seven centuries ago the Greek poet Hesiod composed the *Works and Days*, a rambling didactic manual on what would and would not work in family farming. In that spirit of practical advice, in homage to Hesiod, we can then begin with what the mountain should have been.

California farmland should be flat, not hilly or even gently sloped. Farm tabletop ground because it allows for easy and cheap flood irrigation without erosion. Drip irrigation is a wondrous invention for hills. It is not the promised panacea for the wasteful irrigation habits of American farmers. The city dweller who has a hundred feet of black drip hose in his garden and has been written up in the garden section of his local newspaper loves to lecture the rare farmer he meets on agriculture's immoderate watering habits. At least one suburban tyro has advised me (in all seriousness) on the need to employ her own tiny 1/100 acre trickle system, magnified a ten thousandfold, on our one-hundred-acre vineyard.

You meet them everywhere, these backyard suburban know-it-alls, these New Age gardeners utterly ignorant of the concept of scale. A silly American succeeds at not killing his tomato plant, and so he imagines he can farm 100,000 of them the same way, with no realization that whatever he fertilizes, sprays, waters, picks, or weeds must be repeated identically 100,000 times over if he be a real farmer. Kill one plant and all fall in identical fashion, setting off the chain reaction of destruction that hits everything from your wife's car to your children's lunch pail on the slow but fatal way to the bank.

No, the valley farmer is better off making his land flat for canal water. Do that, and there is no need for $200,000 of filters, hoses, timers, emitters, pressure regulators, and the entire accompanying freight of drip catastrophes—trashy berms, broken laterals, sun-baked and brittle plastic, clogged emitters, gopher-bitten, squirrel-bitten, coyote-bitten hoses, and worst of all, the ubiquitous salesman who incessantly tries to sell you the new, improved, and more expensive replacement system. Wim van Galder, the perfidious Dutchman, distributor for an Italian drip company, used to hound me on the primitive nature of the emitters and filters he had just sold the neighbors the year before. He even ridiculed us for once giving in to him to buy the now obsolete. Often as I nodded off, he, without shame over his last year's sales performance, went on to more germane topics like American criminality, illiteracy, and our generic stupidity. Despite his European university degree and his title "system designer" rather than "salesman," I never bought his abstract gibberish on the theory of pressurized liquids and filtration-resistance quotients. Whenever we expanded or fixed the drip system up on the hill above the pond, we did it ourselves without elaborate contraptions and formulas, without Wim van Galder.

Remember, farmers, there exist skilled men of a different, a more brutal sort than Mr. Wim van Galder. Their entire lives are dedicated to the goal of zero gradients, these big-gutted slobs, like Burt Connely or Armand Cutler, who drive their colossal laser-guided scrapers throughout the San Joaquin Valley to rearrange thousands of cubic yards, to make your land absolutely flat, a two-inch fall every half mile. They are not drip salesmen at all. They talk, not of volumes and pressures, of engineering and computers, but of horsepower and the dirt that their huge machines can carry off. Their accent is not of Europe but still has a strong trace of Oklahoma and Arkansas. They are harsh and often dishonest creatures, but if monitored and reined in for a week, they become miracle workers. When their dozers and scrapers are done, one opens a valve on your farm, and the water travels all the way down the quarter-mile, two-hundred-vine row without interruption, without pooling or breaking. There need be no filters and hoses, no timers or regulators, no maintenance or monitoring, when they are through. When they are finished, you and your children's children will never see, much less need, them and their like again. Their work alone remains—forever. Walk down the alleyway of your vineyard; turn on the hundred valves with your twister; thirty acres of vines are under water in an hour. Rhys Burton, who did not employ them, nevertheless used to marvel at their work, "Why, I tell you, if I had hired Armand Cutler with that diesel scraper of his to flatten out this south side, I could have turned a valve on at one end and had the water reach the other in half an hour. Why, I lost twenty years off my life [he said this at 86] trying to push water over hills with miles of pipeline and standpipes."

The absence of rough terrain also makes cultivation and harvests easier. Tractors do not flip over on the plain to crush their drivers; treaded wheels do not spin, creating enormous scars in the earth; the disk does not rise and sink or bob and weave; pickers do not walk uphill with their buckets full. Without hills there is more chance of soil uniformity and fewer rocks that wear away rubber and steel. Without beautiful hills the flat monotonous soil is soft to the touch, not coarse with pebbles and granite. In an anonymous ancient Greek poem, we read the poignant epigram of the old hillman Euphron, an unfortunate yeoman worn out because he "farmed not the furrowed plain, but shallow soil just able to be scraped by the plough."

The south sixty-acre side of our farm—the oldest, the favorite side—is, as I said, not level as valley orchards and vineyards go. It was the ancient artesian waterhole that drew Louisa Anna Burton in the first place. Thirty acres around the tule pond are the work of Willem and Rhys Burton, small terraced orchards and vineyards that work themselves out of the old pond basin, scraped by horse and mule into tiny fields on the hills. We scrapped Rhys Burton's idiosyncratic system of concrete pipe and put in drip irrigation. Still, it's a mess. Drip lines, pipelines, and alleyways are laced through the two- and three-acre undulating small plots and cannot achieve uniformity, given the terrain. We walk its tree-lined lanes at night to appreciate the terrain, to break the boredom of the level vineyards on the other side of the farm. The unleveled, inefficient contour of the nineteenth century is also where the meager farm profits of the more productive north side go—not to feed us but to sustain its own, its weaker south side and its tule pond terraces.

We grew up with the unquestioned agrarian dictum of my grandfather that "the north must support the south side," never with the more logical, more natural idea that "the north side supports us, screw and sell off the south." That Rhys Burton could not let go of his own grandmother's weaker half, that he would farm his better land to save his worse, is a dangerous legacy in the 1990s, when you must cut off a diseased arm—ancestral land, son, brother, or cousin—to save the body itself. Still, I live on the south side in my great-great-grandmother's house, the original side that was bought from the railroad in the late 1870s, and it's the side I suppose we will try to keep in a year or two when the home place is probably foreclosed, sold off, or rented out as the housing tracts a mile away ooze ever nearer. Still, its hilly, sandy orchards and vineyards, as the old Euphron knew, are where money departs that does not come back.

Farmland should have a house. Homestead residence eliminates costly commuting, hours wasted traveling to and from the fields. A farmhouse also allows the nosy agrarian to watch his ranch around the clock, curbing vandalism, encouraging nighttime and evening chores after the regular working day. Living on the land, the historical central tenet of American homestead legislation, also fosters a rural chauvinism, a notion that your wife's work regimen, your children's desire to go to the mall, your own inclination to hit the bar—all the urban distractions that cost money and waste precious time—are secondary to the salubrity of the farm. Are not they simply obstacles in the way of the no-nonsense farmer, who pumps his own drinking water and stores his own sewage? Pass by the blacksmith's shop, Hesiod wrote twenty-seven centuries ago. Can these unproductive diversions not be clipped out of the working day, every day, when town, not the farm, is a glow on the nighttime horizon? Keep your children away from town, exhausted on the tractor and in the packing shed, arms at work with the shovel, and their perilous voyage between twelve and twenty might pass in tranquillity. Where else can you announce to them in March, "Don't plan anything this summer, you are all in the fruit shed between eight and eight every day." When the Greeks began to live on the land they farmed, the entire history of the Greek city-state, of Western culture, was altered.

Isolation plays a large role in the morality of the farmer. Without a convenient lounge nearby, bereft of chatty neighbors, undiscovered by pretty and younger women, alcoholism, sloth, and fornication are more problematic, are better avoided—even without the positive, supporting network of trees and vines, which exact constant physical exhaustion as the price of their mastery. Isocrates, Aesop, Aelian, and Alciphron all knew of farmers who never went into Athens at all, a whole lifetime in bliss away from the Parthenon and Agora. Have a farmhouse, save your farm and children at the same time.

Seclusion on a homestead that curbs the human appetite for iniquity and vice is not agriculture's only draw, the sole magnet for thousands who still do absurd things to grow food. There are positive enticements as well. (I pass over the perverse delight of the absolute Las Vegas–style gamble, of losing two generations of work or winning five years of income in a single day of rain.) Where else can a modern man obtain near complete freedom at home and work alike? Do you find smug Mr. Hazletown, earl of the local power company, driving a little too fast down your alleyway as he checks your erratic electric pump, raising a cloud of choking, mite-breeding dust on your vines? You say, "Get the hell off my place now, and send someone else out who has a little manners." Do you wish to plant that empty but weak field, to grow apple-pears, perhaps big red plums? No boss says no to you (you wish in retrospect one had); no memo must be countersigned; you counsel none other than your own fears and the specter that you may take food off your kid's table for your failed daring.

Does your hair stick up to fight the comb? Do you have a bad habit of using bizarre four-letter-word compounds, or is your favorite hat soiled and ragged as to be no longer a real cap? Out on your farm you will receive no lecture from your supervisor to clean up your appearance, for you have no director; no coworker will whisper to others that you reek and are to be avoided, for your associates are mute trees and vines and grease-bearing machines that demand that you are soiled and stink for the price of their cooperation.

The work is hard and dirty, you counter, where mayhem and dismemberment are not rare; you are alone where help is distant and compliment rare. Yes, but such toil is never monotony—the real bane of the modern American. You do not stand upright or sit down for hours, for weeks, for months, for years—for life—engaged in rote assembly or married to a plastic, squeaking machine, where the air for the hapless captives inside is stale and the conversation reduced to the varieties of marital infidelity and the comparative value of certain species of garage doors.

Outside your front door you prune for two weeks. You patch wire for a day. You go to the bank in the morning. You hire and fire and rehire monthly your nemeses and friends Max Graham and Juan Gonzalez. You are the local receptacle for the dim-witted and quasi-criminal, your drug-crazed relative Larry Hendrickson and childhood friend Ernie Garza, who show up without warning for a day's picking or digging weeds each week to pay for drugs, beer, or women. I grew up with Delmases, Chesters, Otises, Joes, L. R.s, and Virgils, the Ernie Garzas and Larry Hendricksons of the first fifty years of this century, who work and quit without notice. Anywhere there is a young misfit with benignly criminal instincts, with no belly for the eight to five, with a taste for alcohol and drug, he is the candidate to show up on your farm, outside your front door, to form a lifelong relationship with you of mutual respect and disdain.

But alone you plant Castlebright apricot trees, and rip ground with an offset disk. For dessert you can weld a hydraulic vine-stake pusher or coat your sagging century-old redwood barn with ugly aluminum. And if you feel especially adventurous or silly for an afternoon, you can paint green numbers on your vineyard endposts.

Always you change tasks beside your house, so much so that you wonder how you can manage to grow, tear out, destroy, and fabricate in a single lifetime, so short the day becomes, even in summer when the valley sun sets only in its fourteenth hour. Even though your place be smaller than a city block, you despair how you will ever contour, terrace, level, plant, replant, form, and mold that vast expanse in a single lifetime. How can you come to know fifty thousand vines—more individuals than in an average city—in a mere seventy years? Daily you stuff the brain of your teenage son with wild ramblings about stakes to be replaced, pipeline to be laid, trees to be grafted, in fear that right now you'll keel over and he'll have not a clue to your grand design. You alone, not the power company representative, turn out to be the dust raiser and mite spawner who will race down your alleys, for the life of the farmer is short and so much is to be done so quickly. "My legs, they are sore and they swell, and I can't irrigate as fast as I used to. They have just about done me in." "But you're eighty-six, Grandpa," I proffered in 1976, a week before he went, "and seventy years is a long time to have watered."

Someone rejoins that your workplace and home are not enviable and leave no other room for refuge: "It is very hot in summer and freezing in winter [Hesiod's dreaded month of Lenaion], and there is no central heating, nor forced air conditioning; is it not much different from under the vine, up in the tree, beneath the tractor?" There is no comfort if your mind is wearied and your hands on automatic pilot on an assembly line. But for one with thoughts of planting orchards and grafting vineyards, of restaking vines and roping trees racing hourly in his mind, it is only torrid on the tractor when you enter the house and notice how cool it has suddenly become. Frigidity too is only a past reality once you seek wood for the evening fire: the work is not hot because it is only a little cooler inside, not cold because it is a little bit warmer inside. For the sedulous, engrossed agrarian, the elements, bitter though they become, have no sway. Indeed, they are to be forever welcomed, never shunned. Cold brings dormancy for trees and so ensures they can grow only in your valley. Heat, awful heat, ripens grapes and dries raisins. You—your house too—should be cold and then hot, hot and again cold.

Where else does the flicker woodpecker swoop over your head when you leave the porch and enter your shared vineyard domain? Who at his desk has forty acres of pink peach blossoms blowing in the window, hour upon hour as he writes checks? Who would not trade check, pension, and health care to leave his neon abode for a night-time ride on the tractor behind the garage, rabbit, fox, and great-horned owl paying their due as they race you for a furlong or two?

Office man can get allergies from plaster, carpet, and perfume. Put that same invalid in the middle of a blooming apricot orchard, his arms in constant motion with the shovel, his brain full of calibrations of water and fertilizer, his eyes following pollen-carrying, life-giving bees, and he revives. A great weight has been lifted. His sinuses will inhale dust and pollen—and then clear as the throbbing over the eye disappears. For he will be growing food by himself, not devising an ad campaign for Korean tennis shoes, not typing out the student evaluations of mediocre teachers, and so his mind becomes unconcerned with appearance before peers, with obsequiousness to young men with questionable degrees, with worry about promotion (or fear of termination) to pay for braces and private school. Corporate man has brought himself riches and leisure but not happiness. No wonder Virgil spoke of the always complaining but continually blessed farmer, "O the farmers most fortunate beyond what is due—if they but knew their happiness."

Is not such rehabilitation and therapy, such cure for malady and affection, religious at its heart? For if you see, hear, and smell the incessant birth, death, and renewal of plant and animal, the sudden inexplicable destruction of crop through hail or frost, the devastation of a year's grapes in a second, how can there not be some God that wishes to shake you, to cleanse, launder, and heal you, among tree and vine, to teach that you too—for all your formidable tractor and ferocious disk, for all your grafting and budding of man-made cultivars, for all your effort to defeat weeds and kill off worms—are not unnatural. Exhausted, beaten down, through at last at eighty, the farmer understands he has only been renting his land all along. Man is not the measure of all things. There are things stranger, more wonderful than man. Man is not a *polis* animal. For all his sheds and pipelines, the soil has not been altered, cannot be mutated by the toil of a single man in a single life. That confession of perpetuity and eternity—and his failure to conquer nature—leads to the divine, as the farm itself, its guardian now discovers, was ageless and deathless all the time, put there to inculcate in its custodian that there is something more than interest, equity, and depreciation.

The farm and farmhouse are nothing other than a cathedral of sorts, whose trees, vines, hawks, fish, and dirt itself devour and absorb any who would battle in their company hourly, hourly for a lifetime. After five generations on a small piece of land, the specters and apparitions crowd. You begin to marvel when without warning the momentary insane seizure comes on you, as the vines become blurry in the seance. Suddenly without alert you are made to ponder, who put this locust tree endpost in? Why did "they" plant these hundred-year vines six instead of eight feet apart? A queer concern floats out of nowhere: was it your great-grandfather, his wife, his brother, his son, his grandson who put in the shady and nonutilitarian walnut trees along the alleyway, who dug out the pond, who once bought the wagon rotting up on the hill behind the barn? When you rebuild a shed of Willem Burton and his mother, and notice there are no posts, no beams, no trusses, only eucalyptus trunks, as pristine as when they were first cut one hundred twenty years ago and more, the awful fit comes on again: you petrify and wonder, what kind of men and women were these who wired trees for support and hauled in brick for floors? Belief comes to the farmer, the nature fighter who least of all is seeking it, as religion cures him of many ills and in turn at his end discloses to him that his noble war is lost. The farm, I know, is a far more terrifying mentor than any geegaw found on Sunday morning television.

But there is also the black side to the spiritual isolation of the farm, of living where you work with no escape. The hours alone—no wife, no kids, no friends—finally took Rusty Balakian around the corner. On his last day still he jabbered about chermoya farms, computer-driven solar nurseries, wind-powered irrigation pumps; a *polis* animal beached for life, alone on his dad's forty-acre raisin farm, cursing the oppressive quiet, the shapes flitting between the vines. In a drug-induced craze, babbling with drink, Rusty Balakian put the monoxide pipe to his nose, the knife to his wrists, and the steel-gray revolver to his brain, so determined was he that he not fail, that he of urban promise not be left alone with the smug and triumphant vines.

When you live in a farmhouse on the land, are a few extra hours cultivating the vineyard before your door any different than arranging the furniture within? Who wants a new couch when the vines outside the window are weedy? Better than a new carpet are fields that you can show off, that can make you as proud as a remodeled kitchen does. Where does home improvement begin and farm investment end, your wife complains, when you use only the extra tree-trunk paint to cover the house a dull, cheap white? Who is to complain if your tractor, not your car, is in the garage? In contrast, a city-living man may have a trashy, eight-to-five vineyard outside of town, but none of his suburban neighbors would know it from his deep blue-green dichondra lawn and his scrubbed concrete driveway. He wears slacks tucked into low boots; he even has plastic pen holders in his pockets, this man who can rake his carpet and farm his lawn but cannot grow grapes.

The Greeks invented this idea that each citizen would live and work on a uniformly sized plot. From the symbiotic creed of farm and house grew the entire notion of Western civilization: the idea of a city-state where each citizen had his own square in the patchwork of the countryside, his slot equidistant in the grid of the phalanx, his identical seat in the parquet of the assembly hall. Born out of that environment, the egalitarian community of Greek yeomen developed peculiar ideas seen nowhere else at that time, weird thoughts like constitutional government, private land ownership, free enterprise, and citizen-controlled militias.

When every American carried that grid in his brain, when all of us wanted from the earth about the same as everyone else through the strength of our arms and the bone in our back, we used to be a lot more Greek than Roman. Now assign to an undergraduate humanities class old Hesiod's *Works and Days*, his tale of an agrarian community coming alive in early Greece, and they doze and snore. Then try Petronius's *Satyricon*. The youngsters show a modicum of interest, finding an ancient Michael Jackson or Madonna on nearly every page.

If just 10 percent of our population lived on farms, did not move, never divorced, did not change jobs, and set the parameters of their day by dawn and dusk, the current madness could be stopped. Yet we lack that prerequisite reservoir of agrarians who might still arrest the itinerary of our present culture, of growing shiftlessness, criminality, and material banality.

Oddly, we sheep who follow only fashion and fad still admire the oddball and nonconformist but with legitimate reservations and precautions. The independent trucker is the stuff of pop ballad, yet we still wince at his oil and grease and the petty criminality of the industry. We are nostalgic over religious dropouts like the Amish but learn their iron-willed agrarianism is powered by the sanction of an authoritarian and unforgiving God. The cowboy on the screen and nineteenth-century military hero lead only to silly buckskin fringe and cowboy hats or peculiar reenactments where grown folk don uniforms and reenact the battle of Gettysburg. We are searching, we Americans of modern material and urban culture, for the epic individual among us who says he has an identity that cannot be bought, rented, or leased, for a man the sociologists label "able to resist being drawn into the orbit of industrial and bureaucratic organization."

Where, we the weak ask ourselves, will be these counterpoints to a national ethos that has left us parched and wanting? *Where* will be the often unpleasant individual, the cratered veteran of a continual, a personal struggle with nature, the cultural dissident who will choose still to go it alone in order to protect his old notion of a community, who will have innate distrust for authoritarianism, large bureaucracy, and urban consensus? *Where* will be the man prerequisite to, the exemplar for, democratic and egalitarian government? The ugly agrarian alone is the now increasingly rare voice that says no to popular tastes, no to the culture of the suburb, no to the gated estate. The noncorporate man can always explain to us of a different brand, an aggressive and materialistic urban stripe, how far adrift we have gone from that ideal. His was the grating voice that might have said finance, insurance, advertising, and law—the great sought-after tetrad of the last decade—were not the real work, the true production, the noble professions of our nation.

What other profession is there now in this country where the individual fights alone against nature, lives where he works, invests hourly for the future, never for the mere present, succeeds or fails on the degree of his own intellect, physical strength, bodily endurance, and sheer nerve? In what other vocation now does an American care so little about his own appearance, about the type of car he is to drive, about the title of the job he is to enjoy, about the status of his associates, but so much more instead about the promptness of his action, the unambiguity of his intent, and the power of his promised word? In what other profession is excellence and character certified by the growth and renewal of impartial plants, not concocted by the alphabet soup of B.A.'s, Ed.D.'s, Ph.D.'s, J.D.'s, A.A.'s, M.A.'s, B.S.'s, M.S.'s, LL.B.'s, D.D.S.'s, M.D.'s, M.S.W.'s, M.F.A.'s, M.T.A.'s, M.P.A.'s, M.P.H.'s, or M.P.T.'s?

Is not the vanishing agrarian the true heir of Western culture? In the spirit of the Greeks, he, nearly alone now in this country, believes that man, like wild species of trees and vines, is feral. As orchards and vineyards are tamed by agriculture, so too culture—law, statute, tradition, and custom—domesticates man, teaches him to become productive, and so forces us all to repress and abandon our innate savagery. The farmer, with his keen mastery of grafting and the union of wild rootstocks with cultured species, does not believe that either human kind or aboriginal fruits are inherently good, that through the abandonment of regimen and exaction they might be released to create again a garden of Eden.

No, agriculture—unlike both the squint-eyed exploiter and blinkered environmentalist alike—with divine sanction conquers a sometimes belligerent and reckless nature, improves upon nature, forces for a time nature to serve man, just as the *polis* takes the killer, the rapist, the thief, and the arsonist out of us all. For the hard-headed farmer, the behavioralist, the social engineer, and the psychologist are unquestionably a dangerous breed, a sophistic brood, these who would now seek to rehabilitate the murderer, to parole the thug, to refashion the sinner, to release the inner good, the innate nobility of man from the harm and damage of oppressive social constraint. For the farmer, their protocols, their bitter broth of social science and pop therapy, make about as much sense as chainsawing down all the limbs of his plum orchard to free the gnarled and wild Marianna rootstock out of the stump, to unleash that feral cultivar to grow naturally without the bridle of the graft, without need of irrigation, and fertilization—to produce nothing but its natural bitter fruits, its tiny, ugly, sour offspring protected by spines and thorns. Throughout the history of Western civilization, a farmhouse has led to agrarian, not merely agricultural, culture, to city-states and towns, not palaces and manors.

Farmland should have a history. Previous crops, even be they annual crops, should be watched and recorded over decades. They are diagnostic criteria, revealing the weaknesses of the land, the spot that is alkaline, sandy, zinc deficient, or hard pan. Virgin land was something for the pioneer of the last century, who consumed his life in tragic trial and error in order that you should not. The farmer grows up with the ancestral ground surveyed in his mind through autopsy and paternal lecture, what the chemistry of the soil can and cannot do, which tree or vine can and cannot be planted. He does not clear and plant new land; he does not fell mammoth trees and pry out mossy boulders; he does not because he is a twentieth-, not a nineteenth-century man.

Farmland, especially vineyard or orchard land, should have proven crops—not strange types of new plums that do not set or innovative species of grapes that only rot. In the quest to find something someone wants to eat, the valley farmer must avoid the feijoa, the chermoya, and the chocolate persimmon. There are reasons why exotics like those are in short supply: they either cannot be grown, or if grown, require you to pay for their cultivation. On my initiative we planted five acres of pineapple guavas once, beautiful, delicious fruit that each November rotted in Ross Lee Ford's cold storage for want of a buyer, harvests that we could not even peddle at the farmers' markets over on the coast. A century of expertise tells you that Thompson seedless grapes, almonds, figs, mid-season plums, peaches, and nectarines all thrive here. Join the rabble; plant them; produce their fruits in enormous quantities; you can survive from the ensuing oversupply and crashed prices if you live off your wife's salary in town.

When I asked the neighbor, the old Kaldarian, if he regretted having only Thompson seedless grapes in the midst of the raisin crash, of the wine glut, of the fresh and juiced fiasco, he muttered only, "My vines go out only after everyone else's are out. But I think I'll make money again once it gets down to that." Did not Hesiod say there were two types of agrarian strife: envy, the oppressive strife "that fosters evil wars and battle, cruel as she is"; and the good strife "that rouses the shiftless man to toil, making a man eager to work when he looks at his rich neighbor, who presses on with his ploughing and planting and the good ordering of his house"? The problem with Kaldarian was that he clearly worshiped the former, "the strife who delights in mischief," "who keeps you from your work as you peep and eavesdrop on the wrangles of the meeting place." Still, unlike me, he knew vines, not guavas, belonged in our ground.

Farmland should have easy access to water, should be located beneath, not amid the mountains, so that melted snow flows downward onto its plain. Farmland sits on the valley floor to capture the snowpack and to draw on the once-huge reserve aquifer in its sand hundreds of feet below the surface. The mountains, in contrast, are solid rock. Water cannot seep beneath their adamantine surface; there are no subterranean lakes, no sand strata below. Mountain water belongs not to the mountain but runs over its rock to be trapped for others in the canals and the fields of the valley beneath. Fifteen feet of Sierra snow can cover no more than a foot of topsoil on solid rock, and that thin veneer will always be bone dry by July. Our ancestors built dams and canals to bring the water down so that they should not have to go up.

Farmland should have predictable weather. The mountains form walls to protect the Great Valley from limb-breaking winds, fruit-rotting rains, and tree-killing snows and frosts. Valley man plants beneath, not on top of his wall, where the granite buffers the weather, tames it before it settles below. Even in summer the peaks of the Sierra appear cloud-banked, shrouded in storms that cannot break their way into the valley. Their alpine tempestuousness is the price of our climatological predictability below. Within those rocks above, it can freeze in May, rain in July, hail in June, and snow from October to April. Farm in the mountains and you learn the whole bizarre regimen—freezing radiators, mud-clogged roads, dead batteries, and frozen pumps—of Easterners and Northerners, not the Mediterranean custom of valley hard-noses, with whom you must compete. If you want to farm in the cold, go to Illinois or Minnesota; the valley is the farmland of temperate Mexicans, Sikhs, Italians, Portuguese, Armenians, and Greeks, who came here to be at home once more in its endless heat.

Farmland should have neighbors and it should be settled. Farmers, however odious they be, can exchange expertise with one another, entertain one another, provoke one another, preventing the occasional madness that rural solitude induces. And a particular type of businessman, the foul but necessary broker of harvesters, builds storage and shipping facilities conveniently near his prey, your vineyards and orchards. Sheds, trucking depots, and cold storages, with names like "Sun Valley," "Harvest Sun," "Sunrise Farms," are found every four to five miles. That way the fruit tycoon can appear suddenly in the nearby farmer's yard to market, ship, and exploit his ripening and overripe fruits. He may sell your fruit on consignment back east for two dollars a box, but he sells your fruit. This Lexus and Mercedes owner has no need of a four-wheel-drive truck. He scorns the long, winding drive uphill; he does not hunt out in the hills what he can have in his backyard.

The more numerous laboring men of the cities also must live near a farm to pick the harvests, service the tractors, gas and oil the machines. They have no money, and so for them a day trip to the mountains is often beyond the endurance of their decrepit autos, an outing they embark on every five years or so when they wish to match their cars against the climb. Arturo Cazares told me he went to Avocado Lake (twenty-five miles distant). Once, when his enormous finned station wagon didn't heat up, he puttered all the way on up to Pine Flat (forty miles away). Cazares was no novice traveler but a veteran of several thousand-mile odysseys through the bowels of Mexico. But for a worker to drive his own car up into the lower Sierra to pick fruit—there was no need, no reason for that.

Agriculture, which seeks ceaselessly to render the natural landscape unnatural, also drives out the centuries-old banes of the farmer: the beautiful cougar, coyote, raccoon, deer, badger, wild pig, bear, and snake. No, the grower now fights only animals of a lesser kind, the ubiquitous and philanthropic squirrel, gopher, rat, possum, and blackbird, who have grown fat and therefore vulnerable from their man-made diet. Tougher, leaner, more noble game, crowded out, starved, shot, and trapped, should have long since abandoned the vineyard and orchard that they knock over, uproot, and break down. Trees and vines are not wild environs at all but cultivated plains that produce towns and cities, not fauna and flora of a past century. Even the insects that devour valley grapes and tree fruit are of a lesser, more manageable size, carefully categorized and targeted by a century of entomological counterattack. They are not the enormous-jawed black beetles, the scorpion, centipede, and cigar worm of the primeval forest that can resist chemical gas and fear no predator. Those belong mummified in your daughter's science-fair project, not alive in your vineyard.

The Sierra mountain, where Mr. Bus Barzagus planted his pear orchard, had none of the above requirements and so, as I have implied, violated the inviolate canons of modern farming. There was *no* house, *no* farming history, *no* reliable crop species, *no* neighboring agrarians, *no* water. There was an hour commute each way, with rocks, oak trees, animals, natives, inclement weather, bears, cattle, deer, and rattlesnakes. End of story. Farming there would not work. A decade would be lost, the project abandoned, the land sold off. The mountain was so aberrant from the ideal form of a valley farm that in some ways it could not be a farm at all. I knew that it would not work, but I also knew Mr. Bus Barzagus, and I remembered I had not yet seen him want or fail.

The entire Barzagus family was well known both to us in particular and to most of the south county in general. They were all teachers—brothers, sisters, father, mother, nieces, nephews, sons, and daughters instructing the families of towns that orbited Fresno. The patriarch, the father of Bus and grandfather of Buster his son, teacher and would-be novelist Zag Barzagus, had been involved in local politics and had hired my mother as his secretary for a time thirty years ago. My aunt had taught the Barzagi at the local junior college; my father knew them there as well and worked a little in the campaign. Our clan, with its teachers too, had known some Barzagus or another since right after World War II.

Both families came to appreciate that education and agriculture are not always symbiotic. Teachers make both bad and good farmers. Good, because they are inquisitive and not at all averse to agricultural innovation; bad, like myself, because they put stock in their (usually irrelevant) education and training and so develop by extension the illusion that they know why and how plants produce food. They are good farmers, then, because at times they see no reason to abide by a stupid archaic rule; more often they are bad because that rule was the product of centuries of trial and error and possessed a richer pedigree than the degree they picked up at the university. The teacher's mind will plant pineapple guavas. Better it is in farming, I've learned, to forget that you ever went to college at all.

All of winter 1984 I pruned and tied vine canes beside José San Esteben, with whom I had gone all through school and who, due to an unfortunate accident that had burned off half his lower face, an eye, and a section of ear, had now bought a small farm with the insurance money. "You went over to Greece for a few years. I got more money than you do now," José reasoned quite correctly. "Why get all that schooling over on the coast? You're pruning just like me," he concluded proudly. Why, indeed? San Esteben knew nothing of the Federal Land Bank. Interest rates, new species of grapes and plums, drip irrigation technology, or innovative trellising systems were without consequence to him. He would never lose money, much less his tiny ranch as long as he kept pruning and did not listen to the foolishness of the formally educated at his side.

I only got to know Bus Barzagus, son of Zag, fourteen years older than me, when I returned home for good in 1980. He had earlier thrown away his Ph.D. dissertation a few pages from the end, quit his English post at Fresno State, and at thirty, without any immediate family history in California farming, quite illogically and without capital, mortgaged his furniture, gone out, and bought a small vineyard. The only difference between our respective careers was my greater degree of timidity, a stubbornness in writing those last few pages and thus a future, though as yet unforeseen, reunion with the academic slothful, and the safer haven of inherited, not purchased, land. I maybe had the rougher road in and greater tolerance of the academic hothouse, but of course the prize for courage went to him for surviving in the real vineyard.

Barzagus's entry into California agriculture was nearly a decade before our friendship, and somehow in those ten years he had parlayed that initial land purchase into another, more highly prized, productive fifty-acre Thompson vineyard, replete with equipment, buildings, and houses for various related Barzagi. His Barzagus compound, moreover, was no more than a mile from our own ranch. With the long family history, his similar experience of an abandoned academic career and subsequent refuge in agriculture, the close physical proximity of the respective ranches, and the mutual disdain for corporations and the hard-nosed alike, we all obviously became friends with Bus. Bus Barzagus was about five feet ten, two hundred and twenty pounds, and unusually strong. He had been quite a quarterback in college. Still, he was no match for the mountain.

In farming the successful grower must have one exceptional characteristic that facilitates the growing of plants, one rare gene that does not desert him, does not move or crumble when the onslaught comes, but serves as a fountainhead for all subordinate skill, a life raft for the decades of ruin. For some it is the love of solitude that can express itself in hours of work, ceaseless activity of the lone wolf that eventually converts plentiful labor to scarce capital by one's sixties or seventies. Time passes these solemn by. Drive by them in the field, and they feel sorry for you for moving on to something exciting, not for themselves for being rooted in the commonplace. Unobserved, they wrinkle and gray, toiling in the vineyard while others watch the evening news or take in a movie. They too eventually win and so are rewarded with a decade of security before the end. I hope that happens someday to my twin brother, Logan the Recluse; he tends to stay out on the ranch all day, in the dark evenings especially, avoiding the houses where people, time wasters all, have been known to visit.

For others, as I have said earlier, it is simple parsimony bordering on greed. The love of the ordeal can be expressed in cutting expenses and robbing neighbors. An absence of charity and clemency becomes an end in itself, and the resulting evil also manifests itself in farming success. It is not the desire for money and what money brings but more the passion not to spend what little accrues. Like the misanthrope old Knemon of Attic comedy, this farming tightwad and crook ends up without brethren: No friends or family can abide his presence. He does know, however, that he will die with his land and with cash on hand. He suspects a few even of the just will break down and concede, "He was no damn good, but he left his family a lot of money in the bank." Vaughn Kaldarian, avoided by all who had known him for more than a half hour, passed on with hundreds of thousands in shoeboxes in his closet. In his last days he could not resist stealing a few persimmons from his neighbors to peddle at the farmer's market.

Similar was Carl Otis, a soot-covered Vulcan, who ran his ancient, inherited farm-welding business. Forty years of overcharging, price gouging, and disdain for agrarian customers had left him dying with quite a pile and no friends. His beaten-down, cursed-at wife behind the grimy counter with their retarded son smiled two years ago to me: "Poor Carl, he's going fast with cancer—bone cancer. Guess we'll have to cash out and sell soon."

Some farmers are again nothing other than oversized muscles. The brute power of arms and thighs can substitute for expensive hired labor or complicated hydraulics. Even their prominent bellies have muscle in them. Pounding in stakes or pruning trees can tire them only after twelve rather than eight hours. When wage or mechanical power becomes too dear, they look to their own flesh and so too find salvation of sorts. While energy runs in their tough bodies, they do not want. Hector Soares, with no capital, with no land, his body battered and worn, his appearance betraying fifty rather than his thirty-eight years, continued to rent vineyard all during the 1980s—and to survive. Despite the old equipment, despite the rental payments, despite his crazy wife, he trudged on because his muscles did not rest, pruning, shoveling, at work on the tractor and the spray rig, always, without interruption, ceaselessly his arms and legs in motion.

If, like Hector Soares, you disk all day, five acres of vineyard an hour, fifty acres by evening, you cannot hear for hours afterwards. You smell for days of diesel smoke, and the vineyard dust goes into the inner ear and upper nose, under the eyelids, and between your buttocks. There are no cabbed, no CD'ed tractors for the tree and vine man. Always in a moving dust cloud, you sense nothing in the vineyard but the broken instrument panel before you on the cracked dashboard. The constant roar of the tractor engine is welcome, for its steady drone, uninterrupted by collision or crash, is proof that your tandem disk behind has not smashed through a subterranean pipeline, taken out vines, stakes, and wire, or clipped an endpost—has not then cost you a full day of repair and your children's new sneakers. It is a cosmos of inches all day long as your one-ton disk misses vines and stakes by six necessary and proper inches. If you are in a hurry to finish, to go back and furrow out the freshly disked ground, then you may have to stand to urinate off the tractor as you move down the row, then you neither clutch nor brake as you whizz around the rows. In August you hose down the tractor to prevent for a brief time the metal seat and levers from burning your arms and hands. Your neck is crooked and stiff from watching the disk to the rear slice, rather than pull off, the vine canes. Your kidneys ache from the steady jolts and shaking, as dirt, noise, temperature, odor, and bumping assault your senses and in unison attack the body, which, unlike the tractor beneath, is not steel and rubber, is not fueled by diesel or cooled by antifreeze.

Those like Hector Soares can also prune thirty acres of vines in a single winter, something we had each attempted but failed to accomplish. To do that, you had to be in the vineyard by seven and stay there until five when it got dark in December, every day, every week, every month until February. We grew to like it; Hector always liked it. He likes it even now.

Still others are mechanical geniuses. Like Buster Barzagus, Jr., they repair and fabricate their steel and motors for the cost of used parts, never buying disk, springtooth, or forklift new, exhibiting nothing but contempt for the petty baron of Main Street who peddles his exorbitantly marked-up spray rigs and tractors to the mechanically ignorant. They do not drink, borrow, or purchase when they can tinker and contrive. They laugh at professionals, the money-grubbing pomposity of Carl Otis, and his huge turn-of-the-century presses and lathes. When they cannot survive by growing food they build, manufacture, and repair for their neighbors. Agriculture is not in their blood; it is incidental, never essential, to their existence on the land. They farm only because the vineyard is the natural, the best locus of strange mechanical gadgetry—grape-cane trimmers, fertilizer drillers, berm sweepers, hydraulic vineyard stake pushers, raisin shakers—a magnet that can draw to them the richest variety of electrical and combustion engines in the world. Although they be farmers, they talk with no others of like kind but welcome in the trucker, brush shredder, well driller, and tractor mechanic, who alone will relate the labyrinth of gears, bearings, cylinders, and pistons.

The list of agrarian essentials—rarely is the drive for naked power and pelf included—is long and controversial, but the winning trait more often, I think, borders on simple endurance, tenacity, and cunning rather than abstract intellectual brilliance. An ample and imaginative brain can often get the farmer into trouble in a way that the unthinking back cannot. The artist, the German professor, the chemist, and the architect, even though born and bred on the farm, even though he be no refined and pampered student of culture, does better to leave his cluttered mind in town. If you must read, put your nose in the Old Testament, not Plato or Boethius.

Even so, cerebral, not muscular, power alone created Bus Barzagus the farmer. His unusual rise from an impoverished academic to possessor of capital in the space of a decade (his three-ton-an-acre, fifty-acre raisin vineyard in the inflation high tide of 1980 was worth $750,000) was, I think in retrospect, accountable to his uncanny understanding of human nature and the general agricultural prosperity of the 1970s, the hated years of Jimmy Carter and the Democrats. Oh, he was an extremely hard worker, skilled in agronomy and wary of using someone else's money. He also seemed to sense the mood of the country when it wished a relief from conservative Republicanism and a hard dollar, in turn when it tired of inflation and liberal expansion. He never borrowed much during a depression and rarely bought during inflation. His father, after all, had been a successful politician, a New Deal Democrat who could navigate in a sea of conservative farmers.

Barzagus bought things—his land, equipment, houses—used and low. He knew when to sell high. That is how most of the small (and legal) fortunes in this country are made, judging when the people want to move, when they want to sit, and so buying in or selling out in the transition when the country makes its appointed but radical shift from deflation to inflation and vice versa. To do that requires singular daring when most are vanquished and restraint when all are fervent. Even Ak Akito, the local pesticide distributor, who had nothing but disdain for the educated in farming (for him education was a practical investment, reserved for smart Oriental kids that might become dentists, X-ray technicians, and orthodontists) *admired* Barzagus, the Barzagus whom he did not at all like. "He says some funny things in here to get me mad, and his vineyard has funny things going on in it, but he's smart and makes money, more than I do farming."

The Nisei Akito was at least wise enough to know that while the inherited wealthy patronized him, that while the hard-nosed were late on their accounts, there were very few like Barzagus who both paid their bills and politely listened as Ak went on about his strange theories and fantasies: the tragic government ban of the malathion-treated grape tray; the American bellicosity that had victimized a peace-loving Japan in 1940; the need to keep farm prices low so real agricultural expertise could at last be detected; the various racial categorizations of good and bad grape pickers (skilled Japanese, followed by industrious Filipinos, childlike Mexicans, then shiftless and dim-witted southern whites, and at last incorrigible blacks). Unlike Barzagus, when I asked Ak about various insects in the grapes, he scoffed, dismissing me with, "Don't look, just spray."

Unfortunately, Barzagus became attached to vines, tractors, and houses and would not sell, though he knew by 1981 they had reached their optimum value below the crash. He lacked the killer instinct of the trader and speculator and so was prone to share his gifts with other agrarians in the frustrated hope that they were not weighed down with similar psychological baggage. He would have liked to see his friends purchase and liquidate, loaning his cunning for them to convert land to cash and back to land. Of course, his friends were his friends because like him, they had no desire to mimic banks and insurance companies, much less to be the prominent rodents of local real estate and development, scurrying to and fro with someone else's food in their mouth. As Barzagus's friends tottered and fell, they could at least retain their self-righteous, even smug, disdain for a lower caste of agriculturist who, as the raisin boom and crash finally revealed, had always been in it only for the money.

Farming is not simply a business. True, there are in agriculture credits and debits. Comprehension of capital and labor, depreciation and insurance, law and inheritance, is essential to modern farming. Canons too of supply and demand, inflation and deflation, should be mastered. But there all similarity ceases. The increasing tendency of the consultant to see farming entirely as an offshoot of investment is blasphemous. The businessman does not battle nature with his hands and back. His product—insurance, finance, steel, plastics, or ceramics—does not depend daily on weather, climate, and the infinite operations of the plant kingdom, and it is not brought forth from earth, water, air, and heat. His sustenance does not rest with a single roll of the dice at the end of the year.

Can he not place a few parameters on the unexpected, for his stock and trade is more often human and thus its behavior usually more predictable and accountable? His family does not live in his office; rarely now did his great great-grandfather own his business. His clothes are not routinely soiled. Is his flesh torn, his life unexpectedly endangered, by the idling tractor mysteriously popped into gear, the unknown electric short in the irrigation pump, the brush shredder that shreds an arm? He would not like at all eight hours by himself with a shovel, without a retinue of suits and secretaries, without a voice mail, fax, pager, or E-mail connection with like kind. He would not like to go on solitary retreat for sixty years. He does not shift in a minute from the world of accounting on his computer to a brutish stand-down with a drunken and threatening Gil McCoy, the itinerant pipe repairman. He does not leave the debate with his broker for an overdue date with the poisonous paraquat agitating in his spray rig. He does not type up his government reports after forking twenty-four tons of subgrade raisins onto a truck and trailer. In turn, unlike corporate man, the farmer does not sell when he should, and he farms when he should not, for his nose is in the soil always, oblivious to the unnatural world that buzzes about. Agrarian man has no identity crisis, for he knows each minute, each second who he is, a breaker of clods, a planter of vines, laborer, inhabitant, custodian of his ancestral land. Never more, nothing less.

Worries over the elements—water, soil, air—are distant from the mind of a company businessman. Does he know at any given time where is north, where west? Do the equinox and a northern storm harbor any significance? Does he ever read the signs of clouds, birds, and plant growth? Is any year—drought, deluge, bounty, or scarcity—in a natural sense different from the past? Do patrimonial voices from the orchard and vines whisper in his ear that the dreaded biblical cycle of ruination is now at hand?

No, profit is usually his sole goal, and the consumption of material goods which profit can bring him and his associates of the present age. When the eight-to-five man is sick, no longer wanted, or dead, he is almost imperceptibly replaced, the law firm, ad agency, insurance group, or university scarcely missing his presence. Indeed, they are often rather pleased that his final salary can be halved by an entry-level clone. While the rare scrapper who reaches the top office may engineer the creation and destruction of cities, the planting and harvesting of enormous forests, even he nevertheless leaves little of an individual imprint upon the landscape. Of most, there will be no salesman, no mid-level manager, no legal counsel fifty years hence who will say: "Arnold Brewster created this division in 1993"; "George Sauer got us into motels in the 1980s"; "Loy Hudson's team argued for the new lounge"; "I learned office consolidation from wonderful old Andy Lion." Not a one will say that.

Mostly it's an anonymous world of success and failure, whose rewards are largely transitory—the age-old quest to acquire, exchange, and deplete. The exclusively material wages in the here and now—vacations, electronics, bedrooms, and baths—are greater. But no nephew or grandson will follow to acknowledge, to appreciate, your split redwood grape stakes, to see that your heavy-duty concrete pipe was for his benefit, that the new shop was built on twelve-inch centers to last for someone else, that the pruning, the irrigating lessons, the childhood lectures on the agrarian insight into human character were time-delayed to activate and serve you in your thirties and forties. Were they not, you finally discover, all designed to keep the primeval chain of agricultural expertise unbroken, to link you silently, hourly, forever with betters no longer here rather than with the bothersome present, the often worse in your midst?

So Barzagus was a real farmer who enjoyed the profession and was therefore to pledge his existence to remaining on the vineyard that he had restored. That was a real handicap for the astute financial mind in 1980s American agriculture. A few farmers, for example, like owners of Roman latifundia, sold their places high in 1980, sat out the crash, and then bought back in from the now-desperate purchasers for a third of the original price five years later. Why sit on your untapped gains, they reasoned? Why endure loss of equity when someone else can be lured in to suffer that you might profit? What does it matter that your farm of five generations might have outsiders on it for a decade? If the foolish buyer with outside capital did survive on your home place, why not buy another as good or better? Why live in your great-great-grandmother's creaky house when you could use that house to move to the exclusive shore of an artificial lake in north Fresno? At the crest of the boom in 1980, a frenzied Sikh immigrant—to whom land alone brought status—pulled me over off the road. "I buy your place right now. Half a million, one million, two million. You say when; I get the suitcase with the money. Today? Tomorrow? Next week?" From Petronius's Trimalchio to J. Paul Getty, the smart money has always been with "everything's for sale; just buy low, sell high."

But for the farmer, everything is *not* for sale. Equity, capital gains, the real estate market, are stuffed to the rear of his cerebrum, filed away back in the dank mush of the agrarian brain along with assessors' reports and zoning protocols. Nor is the yeoman a walking résumé, liable to jump house, family, community for the greater cash and swagger of the cross-country promotion, a constant price tag around his neck, beady-eyed, tongue slithering always one day ahead of the posse. Again, if you understand that, you understand why the farmer is now doomed.

Still, this instinctual knowledge, even if not fully utilized, was invaluable in farming, for it allowed the small farmer of Barzagus's ilk to latch onto the national trend, to realize when to hunker down, when not to expand. Even if Barzagus did not profit from the income-producing half of his innate acumen—the ability to cash in on inflation—he was adroit at employing the more important second half, when to bow before deflation. True, he could not speculate when it behooved him. He could not even follow his own advice to stay still during a depression. But he could cut his operating losses and farm cheaply because he could foresee what was happening in the country. He could still farm and not be quite like the farmer he had been in the past—or at least he thought he could. And so he went up to the Sierra with little capital, bought cheap land, and planted a mountain.

CHAPTER SIX

BRIARS AND THORNS

Vines with No Grapes, Trees Without Plums

I.

Let me resume with our own fated ordeal on my father's forty acres in Lundburg. In 1983 we finally entered table grape production, at one time agriculture's exploitative arm par excellence that had spawned Cesar Chavez and his union's black eagle banners down the 99 freeway in Delano. Our trees and vines were now to be producing, and there was a good chance that this additional harvest would offset the continual losses incurred on our home raisin farm. Besides, we had none of the intrinsic disadvantages that plagued the similarly desperate Mr. Barzagus; his mountain was beautiful, but a mere five acres of our soil could outproduce his entire twenty.

But what a strange, bizarre sort of production this Effin plan of table-grape production was, completely alien to anything we had ever seen in low-input raisin farming or in any other type of tree or vine farming. It almost made you think the union was right for a time about the poisons of table-grape production. To produce enormous bunches with big red berries that were to last six months in cold storage, we could have no rot, no bugs, no mildew, only unnatural, inflated grapes, obscene balloons full of red sweet juice that never aged.

To manufacture artificial sterile grapes of that caliber, the Royal grower needed to torment the vine. Exaction required, first of all, chemicals, for it had been learned by the Effin "team" (their nomenclature, not mine) that the price of those beautiful rare Royal seedless grapes on the winter table was that most of the crop rotted spontaneously on the vine, without cause or reason, each week until harvest.

Perhaps it was genetics. Or was decay the product of underproduction or overproduction? Who knows why they putrefied? Perhaps it was the species' unusually dense canopy of leaves, canes, and tendrils that blocked drying air currents. On the other hand, Bus Barzagus might have been right that it was the tendency of the berries to form compact, tight bunches. Others like my brother Knut reasoned that the grape was simply bred with a tough, nonexpandable skin that could *never* be inflated. So for our part, we sensed that botrytis and mildew, like all plants and animals, preferred particular species for their dinner and that Royals were their steak and ribs. A Thompson seedless vineyard could take a light rain; a wet Royal field would begin rotting in minutes.

In any case, after July we simply counted down our disappearing unharvested crop as the vineyard fermented and myriads of the stinking bunches blew up left and right as you walked down the row: one thousand boxes an acre—more dust and spray needed; eight hundred boxes left an acre—more dust and spray; five hundred boxes left—more dust and spray. Still the Royal grapes fell putrid to the ground, not fresh in the box. We were always waiting until the Effin green hats said we could pick the crop, until their choice bunches were matured big and all red and suitable for the elite tastes on the East Coast in January. But by late September, when Effin said "pick," the war was lost, the break-even point of minimum production per acre vanished, and the overseers had us, the conquered—strip the battlefield for their loot, the rare unscathed bunches that remained among the flotsam. Most of the crop had long ago rotted.

Family farmers, not consumers alone, have always without cessation hated spraying, for they do it themselves among their children, spouses, pets, and yard. We usually farmed raisins without much more than sulfur and cryolite, both more or less natural substances. For our own posterity, as a warning for our line to come, my cousin wrote down in his diary the dosage of our new chemical vineyard, the Effin way to produce big Royal grapes for the consumer's holiday table.

1. For the mildew they mandated fifteen and more sulfurings, ten pounds an acre, once a week from April to the end of July, a dirty, awful job that burned eyes and throat. Better it was to dust at night when you could not see and equipment breakdowns turned dangerous. Then there was less breeze and the still clouds from the duster settled inside the bunches. For insurance against sulfur impotence, four trips followed with the big spray rig through the vineyard raining on wettable Bayleton, the new miracle "systemic" mildew killer that was designed to kill any spores that escaped the fuming sulfur. Each pouch you threw into the tank cost $200. We were advised to mix wettable sulfur into the brew, just in case the new mildew killer didn't work. It didn't, at least on Royals.

2. For the bunch rot, an April dose of fungicide was needed. Take your pick. They were all carcinogens. Then we covered the field again in July and a final big August spray of more chemical antiseptics. Effin's team recommended rotating our brand of botrytis killer so we could get all three varieties of "material" (middlemen don't like the word *poison*; it scares away potential buyers) on the vines. Once the cultured berries got fat and red and thus naturally began to burst their thin skins, to rot and ooze, it was time for reactive, not preventive, strategies (the enormously expensive prophylactic sprays never worked; they could make almost anything living sick except their intended victims, the spores of bunch rot.)

Here we were directed to our second line of bunch-rot defense: copper dust, then dry powdered white and brown fungicides, and as a last resort good old desiccating lime, the kind you use in plaster or mortar. The powders were designed to suck up the juice and stanch the wounds before they oozed and spread, which would draw in the predictable horde of parasitic insects. So it was to be brown copper, covered by white fungicide, covered by brown fungicide, covered by white lime. You picked up a dust-covered berry at harvest, scraped off its white powder—then brown, then white, then brown—and found beneath them all mildew and botrytis on the berry skin, both growing unperturbed amid the chemical veneer. At least chemicals that could not kill their intended victim would surely not be toxic to humankind?

Twice a week from July until harvest this went on. The white dust hid the brown. The brown hid the white, alternating white-red bunches and brown-red until picking, until you were left stuck with the last layer of white dust on the scarlet grape—big red berries with generations of white and brown dust on their skins, ready not to end their life in a few hours in the consumer's mouth but to lie unperturbed for six more months in the arctic atmosphere of the cold storage two miles away. Still the treated bunches rotted on the vine, juice oozing out of that ugly grape, more rot spores germinating through all the fungicides, detritus dripping through all the dust, gnats and fruit flies nesting in the cracks. This is a far cry, we thought, from the way Willem Burton dried grapes into raisins on the dirt.

3. Mildew and botrytis are not insects, and fungicides can do little harm to animate, legged pests that can even more dramatically ruin your Royal seedless grapes. The Effin dose: your normal two dustings of cryolite powder for worm. No problem. Cryolite is, as I said, a natural mined substance, toxic only to the creatures who ingest this white aluminum-like metal. True, some suggested that it eventually collected in the human liver. But to my knowledge accidental blasts from the duster even to the face (so far) never have had any immediate effect. Miticide then ensued. Six pounds an acre were needed in two spray trips until July to kill the minute leaf-devouring spider mites. That sweet-smelling mist burned the skin and itched terribly. For vine hoppers (who left their untidy excrement on the berries) two (now banned) sprays of old-fashioned organophosphates. Thank the Nazis for their discovery. Those poisons could gas and kill any-

thing that moved; a full space suit was needed to put them on. And do not forget Dibrom/Sevin of Bhopal, India, fame. It is the old lethal bag at the back of the barn, to be avoided whenever possible, the blanket killer that started the whole notion of organic farming. Put that in your sulfur machine, and when you are done you can watch the flies and gnats drop off the berries for the next six or seven hours. Two hours of dusting, despite the respirator and the goggles, and there was a tightness in your chest that lasted for hours, with an occasional electrical sparking in the brain. An ice-cold beer after a sweltering dusting of the vineyard was an especially bad move, as if Dibrom and alcohol did not mix. Perhaps it is really true, as the urban legend claimed, that occasional poisoned corpses give off foul and intoxicating odors, flooring medical staffs during autopsies when a vein is tapped or flesh pierced, allowing stored organophosphates to fume and crystallize among those gloved, gowned, and masked who probe and peer too deeply.

When I later took sick with a mysterious case of unending mononucleosis, in paranoid fevers I thought again about the Effin plan and those toxic showers we once underwent to produce their grapes, the arms stained orange for two weeks, the white powder stuck in the hairs of the nose and ear, the rash on the neck, down the back. After five years of the nightly sweats, numbness, memory lapses, and dizziness, I sought out for the aches, fatigue, and ruined brain a chiropractor/acupuncturist in south Fresno. He agreed with me that the swollen glands, sluggish thyroid, and reptilian blood pressure could be vineyard related—Dibrom, Lannate, Sevin, Defend, Princep, Karmex, paraquat, Metasisquox related. I want to believe in the quack and his failed vitamin counterassault to this day, his cracking of the back to relieve the accumulated toxins, his hot needles to end mysterious unending fatigue. But Hector Soares scoffed when, defeated, I vacated the vineyard, beached in the house for years with swollen glands, the plug all but pulled. "Had mono once as a kid; everyone gets over it in a month or two. You better start pruning vines again and you can work it out. Otherwise it's your own problem."

In all fairness, Effin did his chemical part too at his plant but predictably with much more efficiency and sterility than we tyros in the field. Once the boxes reached his cold storage they were periodically fumigated all over again—grapes, wood, and paper together—in a cargo container that doubled as a sulfur dioxide gas chamber. Methyl bromide chambers were on standby, ready for any export orders—methyl bromide, the odorless, tasteless, invisible gas that could turn the human liver and stomach into jelly; methyl bromide, whose canisters listed hundreds of precautions, hundreds of ways to die from the product inside. Unfortunately, even these poisons did not satiate the team. Effin fieldmen grumbled to us about the federal government recklessly outlawing the old chemically treated paper box pad. That had proved a wonderful, near-automatic way to kill fungi. Now it was taken away by the do-gooders. There was some talk of a replacement dip, a chemical bath where bunches might be submerged in toxin before storage. It was just talk. We nodded in agreement, worried along with the green hats that the Royals could only be gassed periodically in the cold storage, robbed of a permanent chemical pad to preserve them in their closed box with medicines throughout the winter.

4. Vineyard weeds were no problem. The normal winter preemergent and postemergent herbicide, touched up by a contact plant killer two or three times a season was sufficient, if bolstered by an occasional systemic root exploder. Paraquat, the "contact" weed burner, was the chief worry. Like Sevin, Dibrom, Defend, Thiodan, and Lannate, it too was to be shunned as the poison of last resort. A broken hose, and you hoped the pressurized stream of death hit your arm and not your mouth or eyes. It is difficult doing repairs in the middle of a vineyard with steamed-up visor and awkward slippery rubber gloves, and so sometimes you don't. Better to take your chances with exposed but dexterous fingers. Dead flies stuck everywhere to your tractor; birds died if they drank from puddles where the poison dripped. The manufacturer wisely added a horrible smell and a vile dark color to the brew to ward off either animals or children. Black Death we called it.

Be careful, you millions of urban dwellers who live away from the land, when you advocate social justice and ecological balance in agriculture, you who worry over the scarring of the earth, the pollution of the atmosphere, the befouling of water, the extermination of the wild, and the exhaustion of the poor. There are, after all, only three ways to eliminate weeds, and so allow each farmer to produce enough pretty cheap foods for two hundred other off-farm mouths. If you do not use the carbon-monoxide-spewing tractor and its particulant-raising disk, then you turn to the agricultural chemist and his sinister progeny of cancer-causing, water-polluting, animal-killing poisons. If you prefer that the farmer neither erode, pollute, nor poison to produce your foods at the cost you demand, then the third and last alternative of cultivation lies on the back of the dispossessed, the poor man's environmentally correct muscles that must shovel and hoe what can be more easily disked under or sprayed dead. "Dare you tell me," one San Francisco acquaintance once pressed me, "that it is to be either pollution/erosion/compaction/contamination or exploitation?" "Or a little of all," I proffered.

5. For birds (who loved the easy-to-spot, sweet red grape berries) there were three or four ludicrous solutions. Take your choice here, too: awful, randomly exploding noise guns to ruin your nerves, the choice of Mr. Barzagus, who advised plastic netting, or strips of colored plastic laced over the vine to reflect light into the birds eyes as they ate your grapes. Some growers preferred more brown copper dust: at least one brand paid the predators back with a metallic bellyache as they ate, defecated, and regurgitated your year's work. The prior generations' indiscriminate bait—grain and fruits laced with cyanide that poisoned entire flocks—had long since been banned.

Without cash, we opted for Effin's choice, and that was put down as another debit on the company ledger to be paid when the grapes were delivered. The reader beware here: the following description again is not fiction, much less satire, but simply a small part of what I saw, experienced, and did, not what I heard secondhand or read about. We bought from Effin an enormous red helium weather balloon. We purchased metal canisters of the gas to send the meteorological bag up on a large rope over our vineyard. Then we, who were nearly broke and could not afford to repair family cars or eat a meal out, we who were counting every penny on our losing raisin ranch, bought an expensive patented Effin plastic chicken-hawk kite and, per their instructions, strung it below the balloon to terrify starlings off our grapes. I was ashamed to think what Rhys or Willem Burton would have thought of our nonsense, or worse, what Hans Hendrickson would make of his grandsons' plaything on his ancestral ground. The helium oddity rose, sank, crashed, daily over our field, the attached kite usually tangling up in the rope, the entire apparatus caught and then crippled in vine stakes, canes, and wire below. To most passersby we were untangling our children's weather project or engaged in some community service stunt on their behalf that made good target practice for their own kids' BB guns.

I didn't feel as bad when I saw a few other vineyard fools with red balls floating above their Royal ranches. Like the foxes of Aesop's fable who were convinced to cut off their tails, an occasional red sphere down the road eased the despair. I would see crows and blackbirds play with our kite, circle its balloon, and feed below it. Did not, I argued to an Effin fieldman, bird droppings on the kite prove that starlings could hover, perhaps even perch there and, like canaries, enjoy the ride from the swinging balloon? Still, we pulled it down each night, pumped it up with helium, and sent it back up in the morning, the green hats nodding approval and peddling more replacement balloons, gas canisters, and kites as the situation demanded. The birds nested in the safety of the vineyard each night to be near both their breakfast and dinner. Neighborhood rifle flak and angry winged kamikazes took their toll on the balloons. There was always a green hat to reassure us that we "were saving two hundred boxes of grapes per acre" and that "it had paid for itself in the first hour." Besides, we felt the fool's pride in the ostentatious balloon bobbing above the vineyard, as if we were conducting experiments on the cutting edge of agricultural technology.

The last time the last balloon went down about a mile away, some delinquent member of a demented biker clan who lived nearby drew on it with a black marker, writing "fuck you," among other foul things. We were impressed by his primitive entwined phallus, anus, breasts, and vagina, all figures that attempted to incorporate the natural curvature of the balloon.

Who could send that mess up over your vineyard? Quietly the red plastic carcass was thrown in the barn with all the other Effin souvenirs. I noticed the starling problem eased a little once the gaudy beacon disappeared that had pointed the way to the grapes. Logan, with his graduate degree in biology, argued that the exhausted birds were either sick from the rot, the chemicals—or overeating. About the last day the kite was up, I had seen a large golden eagle, rare in the Valley, drawn to the balloon, fascinated by its plastic hawk below. I could not read portents and the signs of birds, but I knew enough from reading Greek literature, Homer, Pindar, and Sophocles particularly, to realize this meant it was time to put the balloon away. I hoped only that the enormous winged creature was not some divinely inspired warning about the disaster to befall.

So once the vines were in the ground and then in production, we had no choice but to borrow more money on the land, don inhalers, and go to war against insects, fungi, and birds. Green hats came around daily, lashing us on, commiserating over our old equipment, our numerous breakdowns, and the unending, unstoppable rot. The notion that we had planted this perverse vineyard to make large amounts of money seemed to give them added license, as if the raisin yeomen of old now had forfeited the moral high ground and climbed down into the chemical muck with them. (We had.) This was no Rhys Burton south-side tule pond. This was no low-input raisin farming of the Willem Burton type. We were now tiny crustaceans in the corporate farming pool, pouring on the chemicals, trying to produce shiny but polluted fruit, surrounded by a cadre of head nodders and note takers with queer contraptions like sugar testers, scales, magnifying glasses, and litmus papers in their company trucks.

My older brother, Knut, to his credit, gave them the tooth whenever they appeared. "Screw you," "get a life," "give it up," he muttered to Bob Forrest and especially to sandy-haired Dwayne Burchell, the worst of the green hats. He's the one who said I needed a sugar tester ($180) on my belt and tried to peddle me one of the grape-leafed belt buckles ($50). He took prize bunches for his wife and kids one evening on his way home. I was not so openly contrary as Knut but sought a more cowardly form of retribution. I caught Burchell gathering grapes (two boxes' worth) in the field but timidly begged off by warning him to wash them before his kids ate them. But the next day I did take revenge against him with the tractor when he suddenly appeared in front of me in the jungle of the row, emerging from an inspection under the vine. I could only nip his rear with the treaded rear tire and tear off a pocket on his Levis, then immediately jump off the tractor to offer profuse apologies. I had aimed for a more substantial thump to the arm or leg in hopes the tire rubber might at least snap a tibia or crack a wrist.

A crooked insurance salesman from a Fresno office also stopped by once or twice as comic relief to balance out the Effin men as soon as he sniffed out the rot and saw the gnats. "Gotta royal rotter on your hands, eh? You should have seen me about protecting those Royals." (Paperwork could do what tons of chemicals could not?) "Just try to get one year of decent production, then we can get you a perennial policy for your next years' rot. That way we'll pay you for the rot every year. Got a lot of Royal growers on the Guardian plan." He too now is out of business, his table-grape insurance office and associates either arrested, paroled, or bankrupt, his mother company gone back east. But who knows under what guise those of his ilk resurface? Raisin insurance? Frost protection? Living trusts?

Even seedier but deserving of some admiration was "El Sleezo," Herman Cantua—a sinister-looking Mexican national with a pencil-thin mustache, the exact replica of an aging Gilbert Roland—who was making a little money scavenging grapes to peddle in the fall farmer's markets down in L.A. He dickered with us nightly over buying boxes of Royals on the vine, then always later stole three or four lugs (along with the wooden boxes) to top off the sale, desperately trying to raise enough cash to save his twenty-acre raisin ranch down the road. The Effin team despised him and ordered us to order him off. But he was hard-working and paid in cash for grapes, something Hugh Effin never did. Cantua's thefts were in single not quadruple digits, and so it was too difficult to haggle with him when you had already been drawn and quartered by the Effini. The Sleaze also cut out rather than complained about, much less threw down, the rot. Dishonest, cruel to his underlings, physically repulsive with gold teeth, tattoos, and jagged facial scars, he was nevertheless a clever entrepreneur who under different circumstances deserved a white-collar job in the white farm-liquidation business, working for a big insurance or finance corporation that could appreciate his considerable talents and his knowledge of human character. All his trips to L.A. were for naught. He lost his place in the spring.

Orvall Barnes, the local chemical salesman, became a *daily* Royal fixture. We liked Barnes, especially my cousin Rhys, who dealt with him. He never pushed the liquid and powdered death on us but simply smiled (for he knew the chemical bonanza of a Royal vineyard) and said he had a little something in his warehouse that would cure whatever problem came up. When pressed, he rattled off application strategies and chemical components, memorized from company brochures. Expensive bags, sacks, drums, boxes, bottles, tanks of chemicals—all beautifully designed, with bright and alluring colors, many attesting to their innocuousness—soon arrived and piled up in the barn. At least mosquitoes and ants weren't too bad near the stack. For all the slick lettering and labels on these packets of death, a tiny good old skull and crossbones still appeared. I liked that. There was at least some honesty there, some truth in labeling left in this country; the skeleton implied death, and you most certainly would die if you ingested its contents. The powder that fell and the liquid that spilled left an awful smelling mess in the dirt. The more that I, who grew up with DDT and parathion, worked with those poisons, the more I began to think that you might die even if you did *not* ingest their contents.

There was also normal cultivation in the Royal vineyard—disking the ground between the vine rows, furrowing the ground, irrigating the ground, then mowing and weeding the ground. Abnormal cultivation followed, specifically designed, in the historic spirit of agriculture, to alter the natural physiology of the vine. Pruning the vines, thinning the leaves, thinning the bunches, thinning the berries, rethinning the leaves, throwing the canes over the wires, picking the fruit, cleaning the bunches, packing the fruit—thirty big bunches, fifty tiny bunches per vine, six hundred vines per acre. Even the neurotic Theophrastus and Columella, two millennia ago the fathers of modern viticultural science, had not heard of all the things one could do to a Royal vineyard. Crews of ten men no sooner made it through the field than we ordered them back on yet another task. Later, I wasn't surprised to read in the university extension's newsletter that Royal seedless was the most labor-intensive grape ever created. That meant the most costly grape ever created. Payment for each job done by the Effin crews required so many boxes of grapes to be sold. Thin your leaves—that cost the equivalent of four hundred boxes packed out. Thin your bunches—another six hundred packouts. But the chemicals, the balloon, the picking, packing, trucking, storing, cooling, gassing, selling, and distributing had already claimed all the boxes that poor, rotting vineyard could produce. We had nothing left for Arlin Nelson over at the land bank.

At harvest came the boxes with green-leafed logos, containing little plastic bags (with red bunches stamped on) for the grapes to be placed in: thirteen bagged bunches to an Effin box, sixty boxes to a pallet, ten pallets to a truck. Our September was consumed with chemically and mechanically separating the rot from the grapes, placing the harvest into Effin boxes, and turning the entire conglomeration over to his compound for its six-month stay until they either rotted inside a couple of miles from the vineyard or someone on the East Coast bought what was left. We prayed no one died eating them, we really did, even though the real danger was to us who mixed these chemicals in their pure, most lethal state before being diluted a thousand times with water. Still, the weird eagle circling in the sky and the fact that even El Sleezo himself, physically indestructible and a great believer in chemicals, complained about all the dust on the grapes, made everyone but the Effins nervous. "Better dusty than stinking," Burchell told us.

In the meantime our borrowed money paid for all that, even for the Effin inspections, especially for the Effin wait in their cold storage, most definitely for the Effin broker fee. By December the packed and stored grapes were still no more than two miles away from their place of origin. Then one January morning, the harvest but a bitter memory of four months past, we got a call from the snooping Dwayne Burchell. He had spotted a single collapsed berry in one of the stored boxes among the mountains of stored pallets in Effin's icy compound! Panic. Pallets were cut open, strapping slashed, box lids unnailed, papers torn away, plastic bags dumped of their bunches. One berry meant there were more somewhere else. Immediately, Burchell broke out our entire lot, thousands of our carefully packed boxes, shipped them to an Effin ranch, repacked them, threw "a few" away, gassed them, and had them all back in his cooler in two days. We lost hundreds of lugs, paid for the repacking charges, paid for the transportation, paid for the additional gas. "Now they'll last until March," he beamed, once more eating every grape he touched, and then darkly warned, "Don't ever skimp on material again. I yelled at you about this over three months ago." Effin had made nearly $4,000 on the sudden repacking crisis, perhaps airfare and more for the annual fall French vacation. We looked with remorse at our petrified red grapes in the boxes. They looked as pristine as the day we had packed them ninety days earlier, almost as if the boxes were coffins of formaldehyded corpses. How could there have been a rotten berry?

You think the holiday price we finally got was at least high? That we profited from the legal though amoral overuse of chemicals? Grapes, after all, are rare in January, rarer in the spring. Not at all was this the case. The price instead was "pooled." Effin's smaller, scraggly bunches on his home company ranches shipped out first in August, our expensive prizes followed last, some *half a year* later. His own earlier harvest costs were cheap, and his grapes went right out on trucks eastward, bringing two to four dollars a box less. His green hats ensured your crop was last to be picked and last to be sold, your boxes—your hazard—pledged to get the big price the next year. Once all the farmers' Effin boxes reached the plant, they were combined so that we all averaged the price received but not the costs, much less the interest accrued. Cold storage, gassing, box, bag, and broker fees, took almost half the sale price, picking and packing another third, leaving only enough to cover some of the thinning, spraying, irrigating, weeding, and tractor work and, of course, none to feed self and family, much less to pay back the original investment of vines, pipeline, stakes, wires, and three years of preproduction work or to give anything to Arlin Nelson at the bank.

I suppose the only profit in Effin Royals was to pick early before the grape swelled and rotted, to keep them out of cold storage, to sell quickly to recover your expenses, and then to wait for ignorant comrades like us to produce the costly, perfect berry in late September that would allow you to share in their later, higher price. You captains of industry and magnates of commerce can frown on this back-alley con, chuckle at its small stakes and the naïveté of its sheepish agrarian fodder. You can think little of the gullible who fell to Effin, the small-town huckster. But for those dull-headed, blinkered, and satisfied, who care only for growing food and working among like kind, distant from urban dwellers and at all costs apart from modern American man, the implications of Effin duplicity were monumental. They were every bit as important as the mergers and buyouts transpiring back east, for they presaged abject failure for the entire desperate gambit of legions of broke raisin growers who met their own Effins. Indeed, the Effin labyrinth meant eventual foreclosure and a job in town for any mice who could not find their way out. This was punishment that outweighed our crime of bathing the grapes in chemicals.

At such times, trapped in such a tub of rising debt and shame, the yeoman often hears in the vineyard the temptress whispering mellifluously, "Sell it now." Piqued, the farmer suspiciously peeks around to see if cousin or brother have heard the murmurs, evil and cowardly shades all, always there but only a threat, like streptococci, when the host is shaky and compromised. But the baneful ones drift in again now stronger, "Why not sell the ball and chain. Sell, for you have done all you could, on your knees, on your back for nothing, always for others—the broker, the trucker, the chain store, the consumer—never yourself. Depression, war, disease were nothing to what you now face, you who have outtoiled, outsuffered the ancestrals, for you are better than they, harder working than even Rhys Burton himself, than his brothers, than his father and grandparents. It is after all, now yours, not theirs. Sell the land now while you are still young, while there is some equity in the land—your land—sell for a home in town, a year or two of no work. It's your call." With a puff on the cigarette, the paralyzed farmer often squints desperately for an antidote, frantic for a blurred image, anything—a rusting hayrake, horseshoes nailed to the barn, faces in the clouds of bibbed and overalled uncle, bent-backed grandfather, or aunt, the clubfoot, who in truth got nothing in their brief allotment that you might do better. This time saved, now sane, stronger for the provocation, the farmer throws his butt in the dirt, his dirt, but the dirt of others as well. In row 103, at the 123rd vine, alone now the farmer boasts—no whisper, but an awkward and theatrical shout—"Hell, I'd rather crash on the rocks than sell." And so said they all.

But it was also at this point of desperation that a reason suddenly appeared why the Royal bunches rotted and cracked, why all the leaf pulling, berry thinning, weekly watering, daily dusting and spraying, were needed for this "chemically dependent" species. I finally drove over and saw Effin's Royal vines. They were scraggly plants on poor clay soil. The ground was dry, the air dusty. Their bunches, unlike ours, were small and stringy. In short, the vines on these ranches produced a natural Royal bunch, a sweet bunch that was wholly unimpressive in either appearance or size. It did not rot and probably needed few expensive chemicals. It was hardly watered—in contrast to our weekly mandated irrigations that engendered mildew and fungi, which in turn required more spray. Effin raised Royals cheap and dumped them on the market cheap. The Effin plan, the "protocols," were more for others, not himself. The only way to handle this rare seedless monstrosity was to give it slack, to back off on the water and fertilizer, to produce lots of loose bunches with tiny berries, to grow little tiny grapes that no one wanted. Anything else and the grape rebelled. Anything else and the number of berries that rotted and had to be cut out of the bunch made picking wholly uneconomical.

Shame on the American consumer who boasts of his organic preference, his purported uneasiness with chemicals and genetics. At the store he ignores the natural smaller bunch with its bird peck, dull color, and irregular-sized berries, a gnat or two circling in the produce section, like a miner's canary attesting to the safety of the fruit. No, he really wants the colossal, hard, resplendent bunch, with huge, shiny, and perfectly uniform grapes, whatever the costs, whatever the effort, whatever the poison. He lies when he says he wants good fruit, ripe fruit, natural fruit. The teacher, insurance salesman, assembly-line worker, and fireman all want nothing to do with agriculture other than to bring home cheap, pretty, and firm fruit. Many of them, after all, believe in raisin plants.

So in contrast to Effin's unimpressive vines, our vineyard was a seven-foot high, lush rain forest of leaves and grapes. One of our bunches was twice the size of an Effin bunch, our berries were three times his diameters, for we, not he, followed his plan. We were producing an artificial, chemical grape wholly against its own genetic blueprint. Bert Luxhall, the nearby gifted artist, we heard, pulled his Royals out immediately and went entirely organic with other crops. I wondered if he believed that Royals were connected with cancer? Any believer in table-grape chemicals should be required to farm a "chemically dependent" Royal vineyard, preferably next to his house and his own well, preferably himself perched atop the spray rig and duster, preferably his kids in the adjacent yard when the gas goes on the grapes.

In our own defense, we only put up with a year of this Effin plan. Broke, desperate, and unable to pay back the borrowed money, we booted the green hats off the ranch. Without chemicals, without leaf pulling and bunch thinning, without red balloons and plastic chicken hawks, without secondhand anecdotes and adages from Professor

Bart Refino, we would simply make raisins, Royal seedless natural raisins, of an inferior stamp to our Thompson raisins, inferior raisins to add to the doom of our own existing money-losing raisins. We could make raisins until we gathered enough courage to confess our losses and so rip out our beautiful but venomous Royal vineyard. Our wives and parents could keep working to service the growing debt; we would draw about a dollar an hour to keep the whole place watered and weeded and keep up the impressive appearance of the vineyard.

But very soon the disasters that had ruined the raisin industry caught up with table grapes in general and Royals in particular. As I said, finally everything one grew lost money. Finally, Arlin Nelson was right after all. By 1984 we heard the Effin men were suddenly crawling back into the woodwork, returning to accounting or farm consulting. Deflation had ruined the entire table-grape market, even their specialized, esoteric little holiday market. The state table-grape board was not any longer bragging of the price per box received but rather of the total shipped, as if it were still a successful industry when fifty million lugs were dumped at below the cost of production. Could not farmers maintain the industry's worth by producing twice as much and getting half the old price per lug? Advocates of the farmer really believed that it was in our interest to produce in surfeit, to be paid in pittance, as if this new "productivity" and "efficiency" was agriculture's way of aiding the American economy. No longer did slick magazines brag of enormous corporate profits per acre; instead the superfarms boasted of enormous *production* per acre. Again, the D word of depression was never uttered. If we were to be paid nothing, the good farmer, the corporate farmer, could simply double and triple his production, his superior expertise and capital investment in technology and chemicals alone ensuring profitability.

Earlier Flame seedless varieties that were usually picked in August now remained unsold and backed up in storage until September and October, cutting out the first months of the later Royal sales season altogether. When Royals were picked in October and ready to be shipped east, there was a three-month supply of red Flame grapes already residing in every cold storage plant in the Valley. It was like a wreck on the freeway creating a bottleneck, as each ensuing grape variety hit the stalled one in front, the Royals stuck in the traffic at the rear, with the choice between sitting unsold in the refrigerator or rotting unpicked on the vine.

I last saw Bob Forrest, the ramrod of the Effin field men, in 1987 at one of Rollin Buckler's meetings for broke raisin growers. He was no longer playing the role of a sarcastic elder statesman, the terse Henry Fonda aborigine who had seen it all, the confidant green hat lecturing on the necessities of having two metal braces on each vine-stake cross-arm or ridiculing our ten-horsepower, ancient pump with tiresome understated slights like "Gawd boys, get some more horses in your pump to get the water to the vines; they'd like a drink now and then, too," or "I drive Massey, never Deere, wouldn't know how to do any other." Bob loved to walk our field, pick out some enormous red Royal bunch, cut it off, hold it up to the sky, twirl it, find a rotten berry—and then throw all three pounds hard into the ground, splattering it everywhere. Three pounds of sweet grapes sacrificed for a single miscreant berry. "Unless you stop this rot, in a week you'll be doing this all day long," he'd scowl and walk off.

Now at the meeting Bob was sullen and wrinkled up. Effin had laid him off, ended his retirement and medical insurance, taken even his propane-converted pickup. He'd lost all of his retained capital in the raisin cooperative collapse. The wine cooperative upheaval had ceased payment on his scheduled fresh grape reimbursements, and he was about to yank his perenially rotting Royal seedless vineyard. "I'm retired. I don't need any of this monkey business." I drifted off bored, as he followed, puppy-dog like, muttering about the crooks in farming with the sudden zeal of the pathetic convert.

At the same time, the crucial *late* table-grape market of December through March, the key holiday season of the Royal seedless, also vanished along with the early season. Royal sales were now sliced off at both ends. How and why? Under the free-trade policy of the Republican administration that had brought us cheap imported wine and subsidized raisins, there was also a plan to encourage venture capital in South America, especially the new showcase of democracy, Chile. Chile's winters were our summers, and vice versa. In October and November, when the Flame seedless finally petered out, when our chemically stored and primed Royals were finally ready to take their place on the exclusive festive tables of America, freshly picked, cheap red and green grapes from Chile poured into the East Coast and ruined the Royal market. Apparently New Yorkers wanted red grapes picked ten days earlier by cheaper hands down south than five-month-old embalmed grapes from California. The agricultural magazines even bragged about "American capital creating new horizons of farming opportunity in Chile," as their flood of southern grape lugs killed any American grape still unsold by the onset of their arrival. We were told, and I had read, that American viticulturists had gone down to Chile to invest, where banned pesticides and nemicides were legal, where the hourly wage was but a fourth of our cost. Chilean students in American agriculture programs were returning home with computer programs, chemicals, and huge machinery. In short, the Effin method could be done cheaper south in Chile, where statutes about chemicals, such as they were here, where rules governing labor and management, did not exist in their new booming "development mode."

But why, I wondered at this point, do politicians, university professors, and financiers apply to production alone the gospel of global trade, the canon of absolutely free transfer of goods and services, and the creed of cheap, cheaper, and cheapest labor? If grapes can be produced without labor and environmental controls far more economically to the south and thus are to be preferred to American harvests, why cannot we in like manner import academics, lobbyists, bankers, and legislators from South America or Asia? Would they not teach, bribe, steal, and manipulate the law for far less pay than our homegrown brands, at far less cost to the general public? Should not graying tenured professors and smug bureaucrats nearing retirement be out in the street when their seventy thousand a year salary is more than the market will bear? Thousands of fresh Ph.D.'s, millions of eager B.A.'s worldwide, stand ready to fill their shoes at a quarter of their pay to save us the cost of their predecessors' bloated salaries.

In 1989 it all ended; even drying the Royals into raisins made no sense, and Logan and Rhys muttered that the vineyard had taken a decade off their life. I stopped by our ranch on the way to the university the morning we bulldozed the Royals out. They had now been in production for seven money-losing years, and the vines had grown huge in the deep loam, enormous verdant canopies producing large annual tonnages of rotten red raisins. My brothers and cousin had not even told me of the exact moment of their destruction, so much for our tie that had helped spawn those ill-omened vines.

Bus and Buster Barzagus had salvaged the beautiful wooden trellises and put them in their own raisin vineyard. Even now I still like going over to the Barzagus's farm to peer out at the archaeology of our youthful work, especially the endposts and the gaudy, red-painted numbers we had put at the end of each Royal row. In their new home the massive stakes and posts look out of place in the less impressive Thompson vineyard, like battleships on a lake.

The dozer, I heard later, took about ten hours, but after ten minutes I left, confident he would not falter. We would burn the piles and sell the bunged-up wire for scrap, all sixty miles that we had so expertly threaded through all those handmade trellis cross-arms.

I drove away and remembered Isaiah (5:4–5)

> I will tell you what I will do to my vineyard: I will take away the hedge thereof, and it shall be eaten up; and break down the wall thereof, and it shall be trodden down. And I will lay it waste: it shall not be pruned, nor digged; but there shall come up briers and thorns.

II.

The southwest corner of our new Lundburg ranch was planted to Regal Red plums. The trees went in a week after and right beside the Royal grapes. We were busier than I have suggested thus far, alternating the days between raisin farming on the home place and both plum and grape development on the new ranch for the next three years. I once thought this a clever way to decrease overhead cost: that by farming 180 acres instead of 120, we would reduce the expense per acre of insurance, our own draw, and any other built-in expenditure that was aside from cultivation and harvest. In fact, the greater acreage only ran us ragged, ruined our equipment, gave greater exposure to bigger losses, doubled our interest load, depleted our equity, and left us with a new, beautiful, but sterile farm and the pathetic boast that "I farm 180 acres" instead of 120.

Regal Red plums, like Royal seedless grapes, seemed very promising, especially in times of economic depression. They came off early, in May. There were no sprays or chemicals required: a near-organic species perfect for young raisin men who hated pesticides and had fine-tuned their raisin growing to a near-organic level. The fruit was crimson, large, and tasty. It was a Santa Rosa–like plum but with much better color, far earlier and yet larger. A lot of growers were planting it, likewise convinced that it might lead the way out of their own raisin or other assorted quagmires. It required about a fourth of the capital of a Royal seedless vineyard of about the same acreage—there were, after all, no stakes to drill or wire to string and far less stock to plant—and yet it still might pay off the same.

The only problem was, as all unfortunates came to learn after years of failure, that these plums *never* set. "Set," as we will see in the case of Mr. Barzagus's pears, means "set fruit," that is, the blossoms fall off and plums appear in their place. "Not setting" describes blossoms that do not pollinate when they fall off, leaving nothing in their wake for the ensuing year. Instead of ten to thirty tons of fruit per acre, you might harvest in a nonsetting orchard one ton if lucky; instead of ten boxes a tree, one and a half to none. Usually in a nonsetting orchard high labor costs made it too uneconomical to climb a twelve-foot ladder just for one plum a branch. A nonsetting Regal Red plum orchard ends the year in the spring. You simply watch the trees grow the rest of the season, with no recompense at all for the expenses of water and weed control and the embarrassment of growing wood and leaves.

What a unique combination we had created on the best soil in the nation: a stunning vineyard and a towering orchard, side by side, the former producing prolific harvests of rot, the latter almost nonexistent crops of tasty firm plums. Our carefully selected varieties could produce enormous tonnages of fetid grapes and scarcely a crop at all of firm, fresh, unblemished plums. All those years when Hans Hendrickson's cattle and sheep had manured the ground, the eons prior where the floodplain of the Kings River had, Nile-like, fertilized the ground, meant nothing other than verdant canopies of vine tendrils and plum leaves.

Kin looked homicidal when I finally blurted out, "We've lost $100,000 of money we don't have on these awful things, and it's not over yet." Homicidal they were, for how can you walk erect, how can you sleep, when every waking hour for three years has ensured your own doom, freely, of your own volition, without coercion?

We would sink into despair in the fall as the Royals rotted. By spring of the next year, our spirits rose in anticipation that the Regal Reds would at last set. That summer we were in depression from our barren orchard once it had failed again. A month later, in July, we rebounded, buoyed that we had at last figured out how to stop the upcoming grape rot. Over the winter and after the annual grape rot, we still looked for our salvation again from the upcoming spring plum crop. In between the rotting grapes and nonsetting plums we farmed the money-losing raisins. When the Royals and Regal Reds were extinguished in 1989 and the spring-fall roller coaster ended, we were left where we began—with our raisins. I was amazed at our appearance. Suddenly, everyone was not late twenty but mid-thirty and graying, wrinkled, bald, and nicotine addicted. The only humor was black. The small forty-acre farm of parsimonious, mustard-gassed Hans Hendrickson now had an insupportable $180,000 mortgage on it. With the raisin crash in full stride, our boys and girls, all ten of them, were now receiving disjointed, rambling lectures about how they would never be allowed to farm. But like all peasants, ancient and modern, I saw their potential even then, a well-disciplined squad, a small (and free) cadre to pack fruit all summer long in our makeshift packing house. Sweatshops for our own offspring might still save the ranch if only we could find enough capital to start all over with, something new in the post-Royal, post–Regal Red second chance on the Lundburg farm. Arlin Nelson had now retired, smug and satisfied that his word had become flesh. Still, perhaps a new, younger officer might agree with us that there was some equity still left for new tree plantings to rise out of the charcoal ash fields of the torched Royals and Regals.

The new Regal Red plum trees, like the Royal seedless, grew fast in the rich soil, almost too fast, as a few old farmers warned. "They'll grow so quick, they won't put out any fruit wood, and when they do, the plums will fall off." So my uncle, the old defeated plum man (soon to rent us his own orchard and, not a great deal later, to shotgun his own heart), forecast quite accurately. Beside the Royal vineyard, the sixteen hundred surging plum trees made us look like model farmers. Onlookers couldn't decide which were more impressive, the seven-foot high rows of the treelike vines or the acres of lush, bushy, vinelike trees. Sixteen hundred trees meant at a bare minimum seven or eight thousand boxes of plums, early big red plums that might save a raisin farmer, might pay for all the costs of the Royal vineyard, allowing us to enjoy the latter's profits (thirty to sixty thousand dollars?) all to ourselves. I felt sorry for those broke and timid raisin growers who simply sat on their money-losing vineyards with no chance of relief from the tumbling prices, who either would not or could not take fate into their own hands, who did not have the safety net of ground like my father Axel's Lundburg farm.

My brother and cousin, when I was still in school, had heard about this exciting new plum variety and had gone over to Tokuda's nursery (he held the patent) in May 1980 to look at it. About ten model trees were laden with fruit, assurances were given that the first commercial blocks were big successes, but exact locations and production records were not forthcoming. A wise choice, this decision to risk all on sweet, scarlet, and early fruit. Brokers loved the plum.

Supermarkets begged for truckloads. (But was that not alarming as well? Brokers always praised what they could not have.)

Why could they not have it in this, the richest valley in the world? Three years later I quietly went back over to Tokuda's to take a second look at his small block of mature patented Regal Reds, but his nursery was mysteriously closed. I heard that Tak Tokuda had now left the nursery business. Since our trees were in the ground, we ignored the vicious rumors that the new plum was a dud, its creator off into retirement, its growers left with sterile stock. The impressive growth of the new orchard was rebuttal enough.

The farmer alone, I think, has no recourse for his misfortune. He cannot sue the man who sells him plum trees that produce no fruit. I have learned that he cannot litigate when told by the broker that his grapes mysteriously disappeared and rotted in route to Missouri, Virginia, or Maine. He has no recourse when his chemical arsenal sickens, maims, or kills his offspring. The newly installed hydraulic pump is newly installed incorrectly, blowing gaskets and seals. Twenty tons of freshly picked plums rot in the yard, as the repairman begs off, complaining of 108-degree heat, now confessing bewilderment with the antiquated machinery, now impatient with the circling and screaming farmers at his side.

Yet the farmer is indeed liable for the arrogant drunken trespasser who drowns in his canal, for the transgressing dove hunter who blows off a hand in his orchard, for the bank note that is now suddenly to be called in by the new loan officer, for the annual irrigation taxes collected in the years of a drought, for the ruined hydraulic pump that did not lift twenty tons of plums. Year in, year out, he sends his food to be packaged, sold, and resold to the consumer, along the tortuous labyrinth that enriches only those who know nothing of the annual struggle, the elemental fight with soil, water, and living organisms to produce harvests at a profit.

Why is this so? Because, I think, the agrarian is at the frightening pole of this complex society—in return for occupying the bottom, for solitarily challenging nature, all the baggage of litigation and liability flows back down the wire to shock and finally to electrocute those few still left at its source.

A Californian need be no citizen to put his kids in school, have his heart replaced, enlist a state lawyer to sue the state. Most of our country's rural felons dash when their vehicles kill and maim, the police now shrugging, now chuckling, that pursuit of the drunken criminal somewhere on his belly in your vines is but a wasted effort, for would it not be difficult, dangerous, and dirty? At the other extreme, the lawless on top care little now for statute and master plan, scheming cheek-by-jowl with their bureaucratic watchdogs to plant dumps in the midst of vineyard and orchards or to pave over a protected river beach. Valley companies file bankruptcy, real estate sales are fudged, the octogenarian's shaky hand is guided on the will—all under the nose of the regulator and judge, whose cover is the variance, continuance, and exemption.

But the curious American yeoman, he unfortunately is both law-abiding and meager, a lethal combination now in America for the pathetic nineteenth-century relic who will obey the summons, who has not the belly nor cash to scamper and squeeze. So the building inspector, the pollution control officer, the assessor, the zoner, the regulator, can obtain a livelihood from this despised kulak. They can find reason after all for their miserable existence, for their hatred of the independent man, for whom solitude is comfort, never terror. They rather like this plodding agrarian target, so easy to cite, so ripe to ticket and subpoena, the awkward prey who pays promptly for minute infraction and inconsequential breach, who listens patiently, politely, to windy sermon and pedantic citation, stands at attention to government regulation.

An abandoned car littered with loot and liquor may be crashed into the orchard; discarded refuse and tossed poison lap and ooze at the vineyard's edge; bullet holes dot your kitchen window. But the blinkered government regulators pass that mess by to press ahead to you: you are the real scent of public hazard. Is not the farmer's brush pile now too large, an inorganic two-by-four peeping out among the limbs and roots? Are there not diesel spots beneath his ancestral fuel tank? And are his children at work in his barn not yet sixteen? Their ranch inspectors are now at climax; the suited clerks swoon and moan, spotting exposed romex wire and warning labels only in English! Sated, exhausted, the government men relax, asking for a light as they warn of the letter and summons to come.

The kindred consumer in the city who eats his produce also turns out not really to be kindred at all. The farmer's ripe pear or plum orchard by the roadside is looted with impunity by city men in the way that the produce section at the local Safeway is not. "Got these peaches on the way home," the high school teacher laughs to his kids, "the trees were just loaded, some even on the ground; there was really more than they could pick."

Each Sunday the suburbanite prances his horse through your vineyard but, in shock, would need medical attention should you park your Allis-Chalmers tractor and dripping spray rig on his rye-grass lawn. He gasps to learn from *60 Minutes* of microscopic residues on his embalmed fruit but drenches his lawn and shrubs with the nurseryman's arsenal, for the most part chemicals now banned in agriculture. He fornicates, dumps his trash, and drinks outside your back shed but cannot conceive that anyone in a civilized society might do all that on the sidewalk outside his barricaded door. He smirks when he sees your uncombed hair, facial stubble, silly sinking pants and dirty flannel but would be quite depressed and hurt to learn that his own fruit-colored dress shirt, wide pink suspenders, and glaring tie are God's real stigmata, man's true touchstone of buffoonery as well.

The problem with all early producing plums like Mr. Tokuda's Regal Reds is pollination; that is the risk the farmer takes to supply the fruit-starved consumer with the first produce of the year. A plum tree does not want to blossom out in late February and then go right to work producing harvests, having mature fruit ripen on the tree a mere eighty days later. Like the production of fattened Royal grapes, it is not a natural but a human-enhanced phenomenon. Pomologists breed, crossbreed, and mutate species until an early variety whose fruit can set, can be grown, and can be eaten within strict chronological guidelines emerges. More ubiquitous, prolific, and natural mid- and late season varieties that ripen in July or August are the true expression of the species: easy pollination and sets, enormous crops, large fruit, long summer sweetening season—and clogged markets with low prices to compensate for ease of production. Early plums in contrast are either too small, too sour, or simply do not set, the charge both the farmer and consumer pay to have plums when they should not. Regal Reds solved the first two dilemmas, and we thought we could manage the third as well. The late Buddy Messingale, Randy's brother, the once formidable but later obese, drug-ridden, and allergy-afflicted broker, assured us before he died: "You get me that plum to the dock, and I can sell it for whatever you want." (The first was impossible, the second untrue.)

Every Regal tree in the orchard had its northwest limb grafted to a Santa Rosa plum variety. Why not put the needed cross pollen at the tree's beck and call? That way even if the bees were lazy, the wind would scatter the Santa Rosa blossoms all through the Regal Red flowers, cross-fertilizing the plums. I even worried about oversetting and our lack of financing to meet huge thinning bills once the myriads of Regal Reds appeared.

Just in case the wind failed, we ordered extra colonies of expensive beehives, almost eighty boxes set throughout the orchard. I was even determined to—and did—throw rocks at them, if I saw any winged slackers who would not rise and fly out from the boxes for our contracted work. By late February, the entire orchard was a sea of white flowers. Santa Rosa limbs in each tree blossomed at just about the same time, so cross-pollination must have been taking place. Plums must have been created then and there. Bees swarmed everywhere. You could hear them buzzing at work a near quarter mile away. A constant north breeze helped them along, blowing and mixing pollens throughout the orchard. It did not rain. It did not hail. It was sixty-five, well over fifty-five degrees, the minimum temperature that honeybees can work in. It was a beautiful spring that set every other species of plum known to valley tree-fruit men. As we pruned Royal grape canes nearby, we watched each day the Regal Red orchard reach full bloom, white, whiter, whitest.

Associates warned us that the mammoth bloom might bode ill; an overcropping was forecast in all other varieties of plums, and that could cost thousands in thinning bills and result in small, undersized fruit as well as general market depression. It seemed a logical worry: the blossoms were so thick you could scarcely see in the orchard. When the breeze kicked up, it was the same as a light Sierra snowfall, white flowers whirling around everywhere. For insurance we took no chances and so flooded the orchard, and the irrigation water seemed both to prolong and intensify the flowering. I caught each of us feigning work, content just to meander through the orchard, our orchard creation, enjoying its splendor, relieved that the unusual bloom did indeed mean recompense for the days of work on our knees and suspicion at the bank.

What would it look like when the plums set and the leaves sprouted? It would be as green as the vineyard, the shade under the trees almost black in the middle of the orchard, as thousands of tiny plums emerged out of their blossoms. So much for the warnings of the timid, depression-burned elder that these fast-growing trees would not produce fruit. Nothing so hardy, so healthy could play its nurturers false, could turn out sterile. Sixteen hundred trees identically majestic, a mere two that had failed and needed replanting. The local packing houses, the Effins of the tree-fruit industry, sent their field men to lock up contracts as soon as possible. Several saw profits once the fruit set. Yes, we were in a depression. Raisins had collapsed. But even in a depression the wealthy will pay the daring who can provide them juicy red plums in May.

All in the family had confirmed that Rhys Burton had saved our ranch in the Great Depression on a mere two acres of Gower nectarines. When his ninety acres of raisins destroyed him, when the fifteen acres of plums and peaches could not support his wife, three daughters, and the assorted twenty-five relatives who showed up at the Alma Depot (to be ensconced in the barn, perched up in the water tower, laid away in the smokehouse, or cotted on the lawn), his

250 trees of Gowers saved them all. A peculiar milky white Gower nectarine was what the fine hotels in San Francisco for a few years wanted for their breakfast relish, and in 1933 a funny man down in Alma, California, could send them what they wanted each morning and so find just enough salvation in the exchange.

Every summer or so a few of these distant and now affluent brethren from the depression days, in their seventies or more, show up from L.A. or up north in impressive cars, near strangers to me. They hobble right to the door and quite brashly claim they are related to me: cousin, great aunt, or uncle-in-law. Soon these mysterious and anonymous people quite boldly meander about the yard and house. When these aged see on the wall the various pictures of the old man, Mr. Rhys Burton, the patriarch who found them, boarded them, and fed them with his Gower nectarines, they tear up. They say he, a good man, picked them up on the railroad tracks, sheltered them, fed them, and kept them alive within his farm buildings with his tiny orchard of nectarines until the war came and they could go overseas or get into construction. They know more about my great-great grandmother's house, their depression abode, than I, and outside the window they point to a now-anonymous spot in the sprawling Thompson vineyard. They say, "Right there, right there were those Gower nectarines that your grandfather grew and that's what fed us."

This was the ritual for the next seven years: we watered, fertilized, pruned, and cultivated the one thousand six hundred trees. Soon they were over fourteen feet tall and had to be mechanically topped to prevent the huge scaffolds from intertwining with one another, even though the trees were planted twenty feet apart. How could a picker reach all those plums on a twelve-foot ladder if the trees got any higher? They were growing so fast that they threatened to form fifteen acres of one enormous tangled thicket, the frightening, gnarled forests of myth and children's fairy tales.

Anyone who could grow behemoths like that surely could not be faulted for the quirks of nature should the trees turn out to be sterile. More than one arboriculturalist—men who do not give compliments lightly—exclaimed that that was the most stunning orchard he had ever seen. Even a reticent aged grafter with a veteran eye for trunk and limb muttered, "Good soil, good trees—good job".

Herein lay the problem: after the initial year of failure, we did not pull the trees out. How could you when they were perfect in color, size, and shape? Like a doxy enchantress who does not love you, it was hard to let beauty go, to turn the sumptuous orchard into firewood without giving it a chance to put its enormous muscle to work, to apply its formidable powers of growth to setting commensurate crops of plums. Associates were known to tell Axel my father, "Good God, Axel, I have never seen trees grow like that orchard of yours. What are your boys doing to them?" This week, thirteen years after we planted them, I prefer to recall each tree we shoveled in, each year of growth, the majestic canopy and gnarled trunks, never the failed set, empty limbs, much less the lost hours and cash.

Each February we watched the bloom pop out. Bees in response went crazy, and we knew it was finally our year. Regal Reds all through the depression (and to this day, I suppose, if there were any still left) brought twelve to fourteen dollars a box. One year of ten thousand boxes would cover all the money lost on the Royals. Another would pay back the planting costs for both the Royal vineyard and the Regal orchard. A final two harvests would wipe out the entire raisin debt. By the fifth year, we could finally see a profit for all those wrinkles and lost hair, those weeks on our knees before the Effin men. Years six through thirty-five, the life expectancy of a productive fruit orchard in the valley, would ensure college for the next generation, ensure them that they would not farm, and yet provide enough capital so they could at least keep the land. Even Arlin Nelson would again be proved wrong. Young men can substitute labor and acumen for capital. Daring, not timidity, was needed in hard times.

The ultimate stage in the yearly life cycle of the Regal Reds was aborted: fruit fell rather than grew. Small green pointed berries did emerge from the flower, each embryo about the size of a BB. In three or four days a pale but growing patch appeared at its acme. But then on cue the stem of the tiny berry yellowed. The plum ceased expanding and dropped off.

Over the next four or five days millions of small wilted minuscule carcasses covered the ground, and the entire year was over. You walked over the orchard's lifeless offspring, and it made you sick, physically ill. We could not sleep at bloom, could scarcely eat. We could not mention out loud the name Regal Red and finally could not even enter the orchard save to do the minimal amount of watering and cultivation.

Some years a few token plums made it, survived whatever natural pathogen plagued this species. Ten or twenty plums a tree, in place of a thousand or two, usually grew into enormous juice-laden scarlet wonders (these rare survivors had the entire life force of the powerful tree as theirs alone). Picking these artifacts made no economic sense, so usually the war was lost in May, and the rare plums either were eaten by our families or rotted. For the rest of the year, from June to October, we simply watered and cultivated and watched the trees. Without the burden and stress of a crop, the liberated Regals grew unchecked, as fecund in appearance as in reality barren.

But we were not idle, not beaten down at all after that first season of failure. The counterattack was over many fronts, with a wide variety of reserve forces and brilliant *strategemata*. Not one year were we not busy to rectify our initial error and win round two. We divided up the task of finding the holy grail, and the three of us went off in pursuit of the plum's redemption. Pomologists were consulted. Other Regal Red farmers were quizzed. A surprised and "shocked" Tokuda was hunted down and cross-examined. Literature was researched. Strange artificial pollens and sweet mixtures were sprayed into the orchard, either allurements to attract the bees to the blossoms or floral semen itself to inseminate artificially the fruit embryos. We were to do what God would not. Blooming limbs from other species of plums were thrown into the trees—a monumental feat of logistics in itself—at precisely the right moment; their own particular brand of pollen might make for a successful union. New pruning strategies were devised. Water and fertilization, always carefully monitored, were now nuanced and modified according to intricate hypothesis and conjecture. Violent arguments broke out among siblings over compensatory measures and finally over the future of the orchard itself. Still the plums did not set. They *never* set. Even wives and toddlers came to hate the very mention of the word "Regal," hated to wade out into the beautiful orchard itself.

There were near misses. Eight trees planted in the yard next to Rhys's house, placed there as an afterthought to be a hedge from the noise of the road, *set heavily every year*, hundreds, thousands, of enormous, delicious plums. Each of the eight Regal trees tantalized us with the image of what might have been had one thousand six hundred of their counterparts a few feet away only been in similar production. Our rendezvous with the raisin depression would have meant little. The Royal seedless grape fiasco would have been the stuff of satire and anecdote at Thanksgiving and Christmas. People would not have died, I think, nor took sick, much less moved away or looked for work in town, if only the Regal Reds had set and produced fruit—if only they had just set.

Or so I thought. The huge sets of Regal Red plums would have been, I told Bus Barzagus, a dramatic way, a heroic way, to survive the depression, like the Gower nectarines of my grandfather, which fed and clothed the destitute during the Great Depression. I could still have been trucking in tons of plums to Ross Lee Ford's cold storage, not begging bored teens to continue with their declensions and conjugations, if the Regals had just set.

He shook his head. It was a depression. I should know that. Nothing was good. The price for Regal Reds, he pointed out, was high because nobody's orchards set; success would only be the luck of the lotto, a fluke, an artifact not reflective of the general agrarian depression. And even if they did set, the price would then plunge like grapes, raisins, walnuts, persimmons, pomegranates, peaches, nectarines, and everything else the stupid fruit farmer grew in the 1980s. It made little difference, he went on, whether the Regal Reds set and we lost money sending thousands of boxes to the anonymous mouths back east or, like his own pears, they didn't set and we lost money growing wood and leaves.

That truth sounded too much like Arlin Nelson at the land bank. We thought on about the eight trees. Did the constant wind from the cars pollinate the eight trees by the house? Did the warmth from the house provide protection from nighttime chills during bloom? Did the wild beehive nearby in the chimney of the house provide a better, hardier race of pollinators for these eight? Were the eight trees a mutant, a superior strain of Regal Red that had been mistakenly mixed in our original lot? Leads were investigated. Consultations were arranged. Young, confident agricultural extension officers were fetched for their clinical expertise; lame and near-dead octogenarian agrarians were bothered from their rest to mine their decades of pragmatics. Had any seen, heard of, read about, experienced, similar phenomena, kindred species of nonsetting plums from the prior century? Why would eight trees in the yard set, when twenty feet away, one thousand six hundred in the commercial orchard would not? The search for the reason why the plums did not set was as intense as the simultaneous quest to discover why the Royal grapes rotted. In the end, the grapes rotted and the plums did not set. My uncle, veteran of forty plum crops, dryly concluded, "The trees, they're just no damn good, just like your grapes next to them."

The despair, shame, and overwhelming sense of failure, taken root from the money-devouring raisin farming, fertilized by the rotting Royals, now reached fruition with the sterile Regal Reds. What was worse—to borrow money to see your grapes rot every fall or your infant plums drop off each spring? Or to produce both and still lose money? Did it compare with the guaranteed failure of growing $500-a-ton raisins? We debated that conundrum endlessly, as both failed ventures, on top of the raisin fiasco, ended forever the preposterous idea that we four men were all to farm together, that anyone could farm at all. Our farming tetrad "H, H, H, and J," three brothers and a cousin, broke up then and there. Axel alone kept saying, "Everything will work out for you boys if you just stick together."

Worse was the self-incrimination, the sinking feeling that, had we sat on our tiny verandas and done nothing to remedy the raisin holocaust, we might have ridden the hard times out. Consolation came only in self-righteous reassurance that we were Napoleonic men of action, not the whimpering complacent, much less the greedy who desired obscene profits from the growing of food. No, we reasoned, we were agrarians, young yeomen of a venerable tradition who schemed in the fresh fruit lotto only as a last resort to save the gift of generations, to preserve Louisa Anna Burton's tule pond and Willem Burton's rusted hayrake and rotting wagon, to preserve our own families, to preserve ourselves.

The wise, less emotional, and more experienced reader must not, influenced by my dark description, entirely despise us. Please do not conclude that our efforts were solely the futile expression of the ill-informed dilettante; they were not poorly researched, inevitably illogical, much less doomed for lack of muscle or skill. We were not tyros who were alive only because of the hard work and financial acumen of Catherine Burton Hendrickson, my mother. In fact, the hours of academic inquiry, the working days that lasted until darkness and beyond, the perusal of ancient family farm diaries, the insightful economic forecasts and accounting provisions we had taken, all made the failure even worse. Remember, careful research had suggested few species could earn money in an agricultural depression. Regal Red and Royal seedless, of all the fruiting elixirs being peddled by men much worse than the reputable Tokuda and the University of California, alone seemed to have overturned that wisdom, alone might allow all four families to live off the land when others could not and, worse, assured us that we could not as well. We planted in a depression because we thought it better to gamble when most sit tight, to be frenetic when the majority is immobile.

True, for the grapes and plums there was the long wait in cold storage, the rot problem, the inability of setting. But these were obstacles to be overcome by careful agronomy and unmatched toil, all the price of agrarian survival in the midst of a little-noticed depression. Even satanic Dwayne Burchell of the Effin gang perceived the stakes involved in our vineyard and plum gambit. He said once, in both disgust and admiration, as he witnessed the rotting grapes and scanty plums: "I don't know one family in the county where four boys on so little land can still all farm together. Who's dumb idea is that? Who works in town? Anybody got hidden money? How do you get along? Who pays the bills when you go broke? If I had a degree I'm outta here." He blabbered on and on that day, until my brother Knut shut him up. Of course, we thought the lazy, no-good Burchell odd, not ourselves.

How could anyone work in town, much less for anyone else, when there was a piece of land to be farmed? How could any brother (as Knut and I did later) go off to town to search for money when there was a crusade in his midst, when grapes needed to stop rotting and plums needed to set, when labor was required and capital wanting, when life savings and the equity accrued by the dead were pledged to keep producing raisins? I spent most of the years of Royal rotting and dropping Regal Reds trying to defend the earlier undergraduate and graduate years wasted in the failed pursuit of "classics" (1971–1980), not the last nine (1980–1989) in the much greater failure of growing food.

How was I to know that soon the explication of Thucydides and Pliny, the boring exegeses of indirect discourse and subordinate clauses in the subjunctive and optative moods, of all places in the fruit basket of the world, the San Joaquin Valley of California, renowned both for its agricultural expertise and its utter disdain for academic inquiry, famous for the richness of its soil and the bareness of its universities, were to feed my wife and kids from now on, when the best land in the world for trees and vines could not?

By 1985 most Regal Red orchards, like Royal seedless vineyards, were being toppled or grafted over, their owners both furious and humiliated at the same time over their collapsed ventures. For some, a few years of Regal Reds put them right out of business. I once told a neighbor on the home place, an awful agrarian really, that we once grew raisins, Royal seedless grapes, and Regal Red plums; he burst out laughing at my clumsy, overdone, and satirical attempt to caricature abject agrarian stupidity. "How do you live then, boy?" I slunk back, muttering, "By teaching Greek and Latin and some humanities and history." "Ah, that all makes sense now," he guffawed, the implication being that only book learning—and wasted, esoteric book learning at that—could devise such a foolhardy triad as Regals, Royals, and raisins. Yet his turpitude did not make him wrong.

Our beautiful orchard, the biggest, most impressive—and most sterile—in the county, lasted the longest of the local Regal Red disasters, superior stubbornness in maintaining its life now becoming our only sense of agricultural triumph. Sometimes an occasional passerby thought late in the season that the orchard was of another successful plum variety, perhaps yellow Wickson or green Kelsey plums that might at least be shipped to Korea or Taiwan. Under such an impression, he would remark on the orchard's unusual size and impressive appearance, hinting that it must have made us quite satisfied. Then the mere mention "Regal Red" was all that was needed to clear the air, to bring out the condolences instead of the envy by the more humane and polite.

Our orchard too came down, fell the same year as the Royal seedless vineyard went under the dozer's blade, not much before the abandonment of Mr. Barzagus's mountain orchard. In fairness to the one thousand six hundred plum trees, we had three years of unmatched firewood. My brother and cousin cut up all the limbs while I taught during the day. The chainsawing took them all winter, and they harvested cords and cords for the fireplaces, the trees producing astonishing amounts of fuel when they would not bear forth food. It took four fireplaces nearly three years to burn the stockpiled orchard up. When we saw the Regal Red smoke go out the chimney, we laughed that there blew away our kids' college educations. And so it did.

Out in the field, when there was little more than stumps left after the cutting and hauling, the dozer uprooted the remnants, piled and stacked the enormous roots in formidable cairns nearly twenty feet high. Then my cousin rented an enormous propane tank with a long metal barrel, a flamethrower specially designed to incinerate piles of diseased or aged fruit-tree stumps, not an impressive spectacle like ours in its prime. After about two or three days of the gas-fed inferno, the Regal Red roots, like the Royal seedless nearby, were little more than black chips and ashes in the dirt. Ash was all that was left of those tender young saplings and rootings that we once worried would never grow but did grow, all of them beyond our wildest expectations.

So Arlin Nelson was right after all; in but eight years the dozer and the torch had them all. I was relieved that it was foggy all those days of the Regal devastation, concealing our final admission of humiliating failure. Perhaps when the fog cleared, a new orchard could be substituted magically and instantly in place. Logan had heard of a new enormous machine from the cogeneration plant that would come out and gulp down your trees, branches, trunk, and roots in minutes, leaving only a small invincible crater in its wake. No one would notice the difference.

What about those eight Regal Red trees?

They still stand alone in front of the house. They set heavily every year.

CHAPTER SEVEN

THE MOUNTAIN, PART TWO

By 1981 Barzagus sensed the raisin boom was soon to be over, a good two years before its actual demise. He got out of the raisin cooperative and so saved some of his accumulated raisin capital fund before its explosion in the last years of the Larry Black dynasty. (I cannot here chronicle the rise and fall of Larry Black and the damage and hurt of hundreds of gullible raisin growers who believed in the cooperative.) The new Reagan administration boded ill, Barzagus sensed. New plantings ensured eventual oversupply (two tax-sheltered partnerships planned to plant twenty thousand acres of vines alone); consumer demand was not growing commensurably; financial interests were tiring of inflation; and so on. Forget the cheery ag magazines and cooperative newsletter. Raisins were soon to be dead. The challenge for all was then threefold: (1) stay in farming; (2) produce an income; (3) do not borrow, much less lose money. As we have seen, (2) and (3) were nearly impossible, and so (1) was a question of time, or rather, of the extent of vanishing equity.

Over the decade of the eighties, at the precise time Bus was struggling with both his raisins and his newfound mountain above, various peripheral schemes of the Barzagi (his son Buster, the equipment genius, was now at his side, doubling the stakes) were attempted, some successful, some not. Both possessed substantial energy and were apparently not about to watch passively as their raisin ranch took them down. Before going onto the mountain, whose abject failure was matched only by its sheer bravado and daring, I briefly list his lesser endeavors after the great raisin crash of 1983.

(1) In the year 2 A.C. (after crash) Barzagus rented an abandoned Calmyrna fig orchard for the taxes, restored it, and got adequate wages for his efforts: a draw. (2) He also bought a refrigerated truck and peddled fresh fruit on the coast earning wagelike recompense for his considerable trouble: a draw. (3) He devised a no-till, no-frill, no-farm strategy for his vineyard to eliminate almost every cash expense known to the modern farmer: a minus, since the resulting weeds induced Pierce's disease into the vineyard and killed one quarter of his vines, lowering production and requiring extensive replanting. (4) He bought an open lot below market value in the fancy section of north Fresno, and his son used his own labor and farm materials to build a sleek suburban house: a plus; the house, Buster's home, enjoyed a prized location and, like all agriculturally inspired buildings, was extremely well constructed, soon doubling in value. (5) The Barzagi scrounged old electric motors, belts, pulleys, and hydraulica and fabricated an entire raisin stemming, washing, and gleaning line with a daily capability of processing three to four tons of raisins, in quality well above industry standards: a plus; broke farmers and would-be peddlers lined up to have a few of their own raisins stemmed for personal consumption and private sales. The model plant at the local university, used for instructing the future lions of the raisin industry, in comparison was backward and inefficient, its finished raisins lacking the excellent stemmed uniformity of those that emerged from the Barzagan prototype. I engaged that university plant once to stem a few bins of our own off-grade raisins. After six weeks of delay, after phone calls citing the impossibility of the task, after frequent breakdowns and complaints, I took the mashed and cut

remains to be reconditioned by the Barzagi in a few hours; even the new million-dollar stemming machines commissioned by Larry Black over at the Valley Girl Raisin Cooperative now lie rusting behind the plant, an utter failure which left their product smashed, stems embedded inside the fruit.

The formula throughout the decade was the same: deplete no equity, substitute labor for capital, seek deliverance from raisin catastrophe in other crops and processes. Consequently, when Barzagus sniffed impending doom on the grape horizon, the events which then led him to the mountain farm, if not excusable, at least now became understandable: cheap, nearly free land with its owner-carried note; fresh pears, not dry raisins; labor-, not capital-intensive, farming.

Unlike our own failed plans at salvation, the Barzagus design to plant a pear orchard on a mountain involved very little, if any, borrowed money or vanishing equity and was completely self-contained, without dependence on a variety of marketing conglomerates once he ventured into the treacherous world of fresh fruit production. Consequently, when the mountain failed, as it inevitably would given its radical dismissal of all the axioms of successful agricultural science mentioned earlier, Barzagus lost little capital at all. His tab was of a different brand, for he who invests all of his labor and his dreams loses that labor and that desire and in its place gains only the invoice that goes with such sacrifice—physical and mental weariness, accelerated aging, and eventually loss of health itself.

Nevertheless, the innate cunning of Barzagus, his physical stamina, his intellect, the presence of a cloned Barzagus at his side, every bit his match with extraordinary mechanical aptitude thrown in, made the proposition on the mountain in fact feasible in a way impossible for most others. For in the end it was not merely the economy, the climate, the animals, soil, and work that betrayed both; it was in no small part the collapse of the veins and heart of Bus Barzagus himself that said "no more." Years later I caught Barzagus saying as much himself; "I lost a lot of it on the mountain," he coughed out before quickly mumbling a retraction.

The Barzagan plan was simple. Asian pears—sometimes called salad pears, Oriental pears, apple-pears, or pear-apples—were a growing fad among wealthy and discerning American consumers. I saw them as early as 1976 in the lunch sacks of professors at Stanford. They were featured in *Sunset* magazine and the food section of the *San Francisco Chronicle* with enticing lies that went something like "the new delight that is sweeping the country"; "the fruit that every salad ought to have." Crunchy like an apple, they retained the greater and more satisfying flavor of the pear. Twenty different varieties of various sizes, shapes, colors, and tastes, themselves nearly unknown to scientist, farmer, and consumer, suggested that the predictable over-planting of any one species of Asian pear was still far on the horizon.

Besides, domestic interest had pushed consumption and demand, driving prices up to an astounding thirty dollars a box. As late as 1980 there was still not much of a supply. Apple-pears sold for five times that in their native Japan. Should export efforts prove successful, nothing stood in the way of a captive Korean, Taiwanese, and as an ultimate prize, a mainland Chinese market, a billion people dependent on the Mediterranean climate of California to supply them with their out-of-season natural treasure, the apple-pear. California nurseries could not keep up with demand of demanding farmers; no wonder ads announced the waiting list for Asian pear stock was nearly two years. No wonder broke raisin farmers imagined that they could ride out the 1980s on the back of the apple-pear. Even sober, raisin-loving, pesticide-loving Ak Akito pulled out five acres of grapes and planted some. "Its success," he lectured me, "will depend entirely on the right kind of spray."

There were other allurements that piqued the interest of Barzagus. Apple-pears bruised easily. They could not be run over bin dumpers and conveyor belts, probably not even picked into bins themselves. It was a labor-intensive, small farmer's operation all the way—a bucket pick; pails brought right into the shed; unloaded by hand; and hand-placed (the picker-packer wearing soft cotton gloves at all times) in plastic cups inside padded boxes. What corporation would wish to fool with that? In Japan they actually tied protective plastic around individual fruit when it was still on the tree only to wrap each piece with tissue paper when picked. Apple-pears were not almonds and wine grapes that could be planted in endless numbers, mechanically picked, their juice and nuts stored indefinitely in mountains that did not shrink.

You are forced to consider silly things like this in farming now. You must at least ponder the (always fatal) lure of the difficult, the labor-intensive, the problematic species of tree and vine to plant if you don't have a job in town. Otherwise the insurance money buries you; they come in with millions, plant what can be corporatized and mechanized, and destroy the market, pocketing the depreciation and writing the eventual loss off to either government or private investor. The small farmer then must sniff around the margins to track down the more difficult and bothersome tasks unwanted by the more affluent agribusinesses. It was sort of the same logic that draws some into cesspool pumping; I once asked Delbert March of Alma Pumping why he devoted his life to sucking out decomposed feces from others' storage pits: "There's a job for me as long as people crap and don't want any part of it afterwards."

But you interject that the nation's—and the world's—population is climbing at a steady rate. Farmland is disappearing at an astonishing pace. You add that chemicals, irrigation, and fertilizers are taking a considerable toll on the natural landscape. The despairing farmer keeps that equation in his mind to keep him going until next year; that was the mantra of the inflation-rich, resource-scarce 1970s: less land, more people, equals higher prices. But farm compensation in real dollars continues to fall as expenses steadily increase. How did this paradox arise? How does this squeeze linger? Why would not more people and less land save the farmer and his precious commodity of rich soil and agricultural acumen?

In a word, the collision of less land and more people is decades away; for all the alarmist literature of the university, the crunch will come only when most of the present generation of agrarians are dead, their children insurance salesmen and teachers. In the short haul, world populations dodge the bullet because production per acre continues to rise as the last ounce of productivity is squeezed out of each acre through new technology, new genetics, new chemicals. While millions of acres are lost each season to asphalt, houses, and salinity, millions more in South America, Asia, and Africa are cut down, drained, reclaimed, and irrigated for the first time by agribusiness enterprises, throwing their inaugural and transitory harvests on the world market in search of the scarce foreign dollar. The Third World is planted to capture the dollars of the First World, to grow things that we here in America grow, to capture foreign exchange, not to provide staples for their own people.

The world crunch will come, for land really is finite while population is not, for the earth does at last say no to generations of chemicals. The squeeze will come vengeful and cruel; but it will also come too late, I think, for the last cohort of American family farmers who were born this century, who alone would have welcomed the enormity of that epic and last struggle against nature to save their kindred. Asian farmers, like the modern Greeks, farm minuscule plots that are endlessly fragmented and divided; the communist panacea in Russia and China was a disaster. Even the agrarian utopia of France and Italy cannot match the productivity of the medium-sized American yeomen, the most clever and diligent of farmers in the world. Only in America has the mind of the collective and the peasant never taken hold. The yeoman here is distrustful of government, more sickened by the philosophy of the limited good and the moral economy. No, American farmers come alive only when there are genuine shortages, when they are told people need them to produce food, when they are turned loose to pour labor and capital into the ground, guided by their own degree of expertise and daring.

In hopes that their own offspring might also have a chance to grow up agrarian, my grandparents Rhys and Muriel, parents Axel and Catherine, gave their all, deferred expense on themselves, subsidized the money-losing ranch out of their own hide, bought vineyard endposts, wire, stakes, and pipeline instead of cars and home furnishings. They also made this sacrifice because they believed that after their own demise, at last farmland and farming—besieged and surrounded in a sea of houses—would be of some intrinsic value, would provide in a distant age to come some kind of living for their line sheltered from the urban nightmare. Though both generations were not romantics, I think they half expected the yeomen, their progeny, would become honored professionals, like the community doctor or professor of the 1950s, once the nation called upon them to feed and save the world. They were generally right, I think, in their ultimate forecast but well short by one or two generations—and perhaps also they were off the mark in their confident estimation of us.

Trees and vines, remember, are themselves separate genres of agriculture and so in turn draw in an odd breed of farmer. Blunders cannot be ploughed under in the spring and fall. The inferior custodian who overfertilizes or underwaters forfeits more than his yield. His sickly trees take a year to three—or never at all—to convalesce. In the process they continue to reveal his inexperience to both inquisitive neighbor and random passerby alike. The capital of his land, of himself with eye always to the future, never the present, is assessed not merely in the richness or bareness of the soil but in the vigor or infirmity of plants of a generation, whose very welfare is every bit as important as the annual crops they bear. And a vineyard and orchard tie one down in a way that grain and cotton do not; this peculiar farmer plants his stock again not for the year, but for a lifetime—not his alone, but his sons' and grandsons' or any who would assume the orchard's charge in the wake of his inaugural failure. For the planter of trees, doubt and fear abound that what he plants might turn its savage head back in anger on its creator to begrudge the years of its youthful care with a maturity of counterfeit produce. If the plant be a peach of poor hue, a nectarine that will not size, a grape that cannot endure to harvest, an apricot whose very bounty and beauty ensure overplanting and overproduction, a walnut poorly matched to a particular soil type, the investment of the farmer's life wreaks havoc on all of his posterity who would farm it, these trees and vines that must be cultivated, watered, and fertilized for years before they reveal the true nature of their harvests. I have seen lush and hardy orchards and vineyards mysteriously produce either rot or nothing at all, and I have seen meager sticks planted on sand whose appearance belies their power, whose choice plums have sent a whole generation to college.

On occasion the lure of trees and vines—the superior beauty of the orchard and vineyard, the high-stakes gamble of enormous profits and catastrophic losses, the innate desire to leave something behind other than bare ground—infects even the colder-hearted cotton, wheat, and alfalfa man of the valley. After years of success with rice, vegetables, or silage, a few of these row croppers at forty, fifty, even sixty years of age have been known to stake all on the table and so to enter the less lucrative, more prestigious world of the viticulturist and orchard grower. One hundred sixty acres of flat and empty earth can look awfully small. But transform that same ground into a ten-acre pear grove, twenty acres of apricots adjacent, thirty acres and three thousand five hundred trees of late-season plums behind them, in the corner forty acres of raisin grapes, and in the back sixty acres of towering, high-trellised table grapes—irrigation valves, endposts, and alleyways latticed throughout those carefully designed blocks—and that 160 acres becomes a lush jungle, where the farmer can plunge in and become lost, a forest that can live on when its planter cannot, a receptacle where all the brains and experience a man has accumulated can be put on ostentatious and permanent deposit.

So I have seen such queer dreamers for no good reason sell off their huge cotton pickers, their eight-wheeled monstrous Deeres and Internationals to purchase less impressive, more pedestrian tree and vine equipment. I have seen them harvest no field crop for three years as they convert to the world of trees and vines, to worry over acres of minuscule young grape and peach stock, to leave the security of mechanized labor for the complicated thicket of hired pickers, thinners, and packers. I have observed them go from annual profitability on their debt-free but nude ground to enormous mortgages for the rest of their lives on orchards and vineyards that are not needed. This allure of the tree and vine farmer and his litter of thousands of gnarly stumps infects more than the urban refugee, more than the corporate investor. It snares even the crusty, stubble-whiskered cotton man, the solemn melon grower immune from romance and yarn, the cocky tomato master himself who feels suddenly empty when he mows down his crop each fall. If one wished to learn how Hellenic culture is to be distinguished from the Near East, he would not be wrong in saying that the Greeks were planters of trees and vines, Pharaoh's people but wheat and barley farmers of the season.

Barzagus figured that skin and surface sensitivity were only part of the apple-pear's negative attractions. Other obstacles also suggested that investors with vast amounts of outside farm-destroying capital—insurance capital, finance capital, lawyer capital—would be wary of the treacherous little apple-pear and so not plant the enormous agribusiness ranches of one, two, and three-hundred acres once they sniffed high prices, the practice that had been so detrimental to valley viticulture and arboriculture (pistachios, wine grapes, kiwi) in the late seventies.

Apple-pears died of fire blight. That was a horrible bacterial disease that came on during their week of bloom in early April, the AIDS of agriculture that had no cure, that tortured and enervated before destroying its victim. Through some sort of unknown transmission, bacteria entered the petal, burrowed into the limb, and then for the next three months destroyed whole limbs, eventually in a year or two killing the tree itself. It was called fire blight because the first symptoms were entire limbs suddenly blackened as if held to the torch.

Whole rows of thirty mature apple-pear trees in our own orchard, ramrod trunks, beautiful, well-shaped scaffolds, the work of ten years, have become scorched and shriveled black over five days. The only preventive remedy was to spray commercial grade streptomycin or tetramycin (we joked about placing our sneezing young kids in front of the spray rig when they took sick) and then to hope the antibiotic killed the invisible bacteria (it never really did).

Once the bacteria penetrated the tree, the reactive strategy was to go in immediately with saw and pruning shears, amputating limbs a foot before the burn line, and then removing and torching the befouled dead wood at once. Nothing might stop an investor colder than a verdant pear orchard turned black over a few days, an orchard that would not reach good production again until another decade or so had passed. Even the backyard suburbanite could not grow his own apple-pears; I used to get calls from friends in town to come over and explain why their six-year-old trees were now dead and black. Never again, they promised, would they plant a slow-growing, quick-dying pear tree. Rightly the Romans said that he who plants a pear orchard plants it for his grandson.

So, as Barzagus calculated, it was no easy thing to grow high-quality apple-pears in the San Joaquin Valley of California. Tommy Takahashi, our neighbor to the south, told me they'd burn black when he saw our own orchard go in. "They'll look great for five years, then one week they all die, just like mine in 1959." It was hard to figure Takahashi, now in his sixties, cash-laden, debt-free, interest-earning, ever to have been a gamer, who twenty-five years ago in his mad youth played his hand in the apple-pear fiasco. I countered, "Did you have tetramycin, streptomycin, agromycin then?" Tommy ignored the question, "You watch, they'll burn." He was right. They do burn.

Other angles were sought out to ensure the success of the Barzagus mountain ranch, for the stakes were high for this iconoclast who had no desire to watch the raisin crash take everything his hands and back had built, for this misfit who would stay on his feet as he bumped through the maze of agricultural depression. Capital was not wasted in purchasing new bare land. In the valley unplanted tree land could cost by 1981 nearly $6,000 an acre (far more, then, than the soon-to-be price of producing orchard or vineyard itself). Up in the mountains, in contrast, rocky forest ground still went for $1,000 to $1,500. The $4,000 to $5,000 you saved would pay for all the development costs and the farming during the initial four or five years before production. Find cheap land, put in the trees, reap the profits, save your ranch below. I think now the land's cheap price must have first led Barzagus up there.

More importantly, Barzagus, no tree-fruit man himself, sought out a partner skilled in all facets of plum, pear, and peach growing, a relative of sorts by marriage. Farmers still hold this antiquated notion, in times of spiraling divorce rates, stepchildren, and transient lifestyles, that blood and near-blood ties in themselves ensure loyalty, solidarity, and honesty, a nineteenth-century anachronism that causes only heartbreak in the twentieth. Randy "Messy" Messingale (whose sister had married the brother of Barzagus's brother's wife) farmed thirty acres of trees and vines of his own near the Kings River, one of the very few true hills that existed on the valley floor. He was younger than Barzagus, without dependents, conversant in both farming and mechanics, and had even greater a reluctance than Barzagus for borrowing money. Where Barzagus was prudent, Messingale was outright niggardly. Messy was well into his thirties, nearing the age most apt for agricultural success, when youth's energy and strength are tempered by skepticism and caution brought on by age. After working all day in his orchard, Messy used to lift weights in the evening, not to impress the ladies nor to improve his health but simply to increase his efficiency in the lifting, pulling, and pushing of tools involved in his daily agricultural regimen. For all his outward pleasantness Messy occasionally puzzled me; to save the cost of a canteen, he might drink out of an empty herbicide container; to buy a two-dollar part, he could price-hunt at four stores. He once told me that he had it with lunch at McDonald's because they had become too high-priced.

In fairness to Barzagus's selection, Randy Messingale also felt more at home in the mountains than he did on his facsimile by the river. This modern would-be pioneer, this tree cutter, and rock mover, had also spent time in Alaska and was instinctively attracted to the vanishing challenge of reclaiming virgin earth, to working high above and away from people, to battling the rare granite and snakes in order to grow trees. More mundanely, Randy's daily experience on his own small river ranch with rough terrain made him ideally skilled in the creation of drip irrigation systems, familiar with various clay and rocky soils, and a veteran combatant against burrowing rodents and the ensuing cargo of washed-out roads and fields.

But the price of Randy's expertise was the inclusion into the fold of the hard-nosed, someone like a Mohinder Kapoor, who would devour his own to keep on his own land. They do keep their land, the hard-nosed like Messingale. They are often ethical, law-abiding people on their visits to town. They even pay their taxes and stay under the speed limit. Stay clear, however, when they are on farming ground; you're fair game then.

Messingale had a brother, Buddy. He was a fruit broker for the local San Francisco and Los Angeles produce houses, those gateways to the nearest of urban markets that charged a 20 percent commission to handle your fruit. Theoretically, he could be trusted to sell the pears promptly and without the usual artifice (i.e., loads that vanish on the road; fruit that somehow rots in route; tags that are switched and lost; checks that don't arrive in the mail; quotes of nonexistent prices; pallets that shrink; and so on). Barzagus could thus tap in a single family both agricultural and marketing expertise. Messy would figure out how to grow pears in the mountains; Buddy would move them into San Francisco and Los Angeles supermarkets; and they would all congratulate themselves at an occasional family reunion.

But here, too, problems loomed. Buddy was as fat as Randy was lean, as excessively lazy as his sibling was industrious, as carefree and careless as his other was unforgetting and unforgiving, as short-lived as his younger was hale. With all due reverence to the recently departed, I confess that I tolerated even Randy the farmer better than I did Buddy Messingale the broker. Both were small weeds, but there was no chance that the former would show up at your house in a twenty-year old Cadillac, with booze, drugs, and L.A. prostitutes, to talk about fortunes to be made on the illicit side of agriculture. It was only when Buddy did do that and asked about renting a few acres of land for drug cultivation and distribution that I realized that our Santa Rosa plum load of the year before, which he had told us "blew up" in the reefer truck, in fact probably had not. More likely he sold it on the side, and it ended up in the bellies of those in Presidio Heights.

Because of the Regal and Royal challenges, I did not witness Barzagus's daily routine once escrow closed and they all went to work on the land. I only received details about the progress from occasional visits to the mountain and more frequently from Bus himself when he was on the valley floor. But would not a lengthy, necessarily thoroughly depressing description of the ensuing catastrophe be redundant, anticlimactic, given the documented abject deviation from agricultural practice mentioned earlier, given our similar development catastrophe, given the general state of the agricultural economy? The eventual failure, after all, was preordained. I can, then, be summarily brief in chronicling what I think most likely happened on the mountain over the next six to seven years.

The hillside itself was situated in a small hidden recess, a picturesque mountain valley of sorts above the foothills but below the true alpine Sierra. It was about an hour's drive from Barzagus's valley ranch, a route little traveled by any other in his family, for whom the mountain ranch was just that, a faraway abstraction in the hills, not the household's inherited lifeblood that had to be saved at whatever cost. The road, not just the distance, was also a problem. After driving fifty miles of curvy and undulating patched asphalt, the last mile involved a one-lane steep descent into the tiny mountain valley, where the pavement ended. Then an easement allowed a quarter-mile dirt drive through a stream bed. In summer, it was bumpy, dusty, and potholed. From fall to spring it was nearly impassable with mud and fallen debris. If entry with a pickup was problematic, how was a flatbed with ten tons of pears supposed to lumber out the alleyway and up the mountain? But why worry, Barzagus reasoned, do not precious cargoes of scarce fruit always find their way to market? Strangers, with muscle, come out of nowhere to accomplish for you extraordinary feats of logistics in the valley if only they smell profits to be shared in the growing of food.

There were no farmers anywhere near. No Kaldarians, Kapoors, or Takahashis. No mountain veterans were present to add their two cents about frosts and rains, to warn against planting new varieties of pears or apples, to advise about deer and cattle, to argue with over irrigation and pesticide strategies, to check the pump at night or scare off intruders, to display farming technique that might be nuanced, adapted, copied, or rejected entirely. Neighbors are not all problems. Without any interest in your own feelings, the adjacent farmers boast straight out to you where you are wrong, when you lapsed, how you failed, why you will go broke. Because they are not family, much less sensitive friends, you are not coddled. You can learn as you hurt. They are like wizened professors who have standards and so lecture to empty classes. If you can stomach their bitterness and pride, you can learn.

When the immense, sullen neighbor next to our tule pond, Mohinder Kapoor, told us he'd eventually have to buy our ranch from the land bank, we debated and argued to find out what we were doing wrong, what transgression against the agricultural canons had caught the eye of this enormous and enormously successful Sikh, this king of the hard noses. In the end, it was merely that we quit pruning at dark. He finally let on to his rather mundane, silly secret of agrarian triumph, to what had won his ranch but ruined his circulatory system: "You prune your last row in the dark, and then when it's pitch black and you're tired, you must do one more."

Despite his bizarre logic and his love of the predatory jungle, Kapoor was worth listening to once in a while about the dynamics of work. After all, before his first heart attack, all three hundred pounds of him had farmed 250 acres of vines all by himself—a pound for every acre with plenty to spare—no hired man, no occasional tractor driver, no irrigator, no kid to help out. He had done it all alone, eighteen hours a day, thirty days a month, year in and year out. But again, it had claimed half his heart muscle.

And he was not at all a sly water pilferer like Kaldarian. Kapoor stuck his head in your face and said flat out, "I am taking the mountain water, all of it, yours too, boy, until someone—maybe you—on the ditch thinks he can cut me off." Much later I stood down his postoperative ghost with a raised shovel. Armed with wood and steel and a four-valved heart, I could take on well enough what little was left of his once 250-pound frame, for it was water he was stealing, and you care little for a farmer's medical history when he has his hand on your ditch gate.

In place of the Kapoors and Kaldarians, the occasional landowners nearby the Barzagus mountain were nonfarming recluses all, mostly second-generation dust bowlers who liked the nostalgic Ozark-like terrain of the low Sierra, where there was freedom to shoot randomly, beach dead cars in front of the door, and incinerate rather than recycle their trash. No homeowner's association forbade them to have more than three abandoned vehicles in their driveways. They were typically armed and frequently on weekends to be seen on their porches built onto mobile homes. Most were retired cannery workers or forklift drivers from Fresno, who could buy solitude cheaply in the mountains. To be frank, they were up there because they were not happy with the Third-Worldization of California, the recent onslaught of two to three million illegal immigrants from Mexico. Unlike the legions of affluent coast dwellers who were leaving California for the more Anglo and depopulated Oregon, Washington, Montana, and Idaho, the Sierra clans were at least honest and outspoken in their prejudicial discomfort—and without funds to move. Still, many of them were to the eye grim, even alarming folk, who convinced you that the real danger inherent in the multiculturalists' balkanization of America was not strident and whiny minorities but the more numerous sullen and violent pink men such as these bearded fed-up minutemen, shooters, and bombers who rather welcomed an armed counterresponse. They again were not farmers and thus at best of no help to Barzagus; at worst they could become an occasional and dangerous nuisance.

A federally funded and newly built Native American village was a different story. Its expansive three-bedroom houses were very close indeed, and that settlement ensured that there would be plenty of trespassers, hunters, vandals, and thieves. No wonder the purchase price of the mountain was so low, the area mountaineers exclusively armed and lower-caste hillbilly. No Sierra-loving wealthy refugee from Los Angeles would want to encounter there what he had fled from down south. No Bay Area Sierra Club member, however much enamored with the forested spot, worried about the destruction of its aborigines, plant, animal, and human, would want to experience the vestige of that national treasure in the flesh.

The Native American unemployed, at first encounter innocuous and with a noble pedigree, very soon sought distraction in rifling though the goods of someone who might try what they could or would not. After losing batteries, tools, and small equipment, Barzagus was forced to bring all his valuables with him each day up and down the mountain. His equipment, like himself, became a commuter. Anything left overnight was either destroyed or stolen by morning. Gratuitous bullet holes and small fires suggested that, like most of the crime in America today, the damage derived from boredom and petty malice, not real need. They, like most criminals of color now between twelve and thirty, were not especially hungry, unhoused, or poorly clad, at least in global terms. Much less was there any studied political agenda; they had no real plan to loot lifesaving sustenance from an exploiting and paler class. Their criminality and lawlessness were simply the familiar break-in, shoot-it-up, burn-it-down kinds of fun, subsidized by a generous government fellowship.

In fact, there was no urge on their part to tap the mountain for anything—wood, beef, or fruits—when the government gave a three-bedroom house, a recreation center, a monthly check in the mail, and the promise of a small gambling and bingo parlor. Everything the farmer was to bring over the rough terrain in his hour ride was to be taken home at darkness. You own title to the land, but someone else possesses it between nightfall and morning, every day, every week, all year long. Before the actual work began, easy accessibility and security—the few precious givens left in family farming—were lost to Barzagus outright. The mountain was not empty. It was full of the antiagrarian, not merely the disinterested, much less the sympathetic. It was quickly shaping up as no place for a raisin farmer, even if it had been the $1,300-a-ton raisin farmer of the awful Carter years.

The twenty acres were on the slope of a tall hillside. Messy and the Barzagi went right to work the winter after purchase, cutting down magnificent towering oaks and dragging out sharp, protruding boulders. A stream traversed the property, so a bridge and assorted conduits were needed to span the two proposed orchards and to prevent uncontrollable gullies. In the San Joaquin Valley, in contrast, an old orchard or vineyard is dozed under in an afternoon, its carcass burned in a day or two, the land releveled in hours. There it is a continual process of agricultural creation and destruction of what valley man—not nature—has produced.

Not so in the mountains. True, a little land is cleared, some orchard is planted as the trees, stream, animals, and rocks are for a time put down. They are never quite mastered, even by the chainsaw, dozer, conduit, and tractor. You scarcely even rent land from the mountain that the county says you own. Ten years from now, twenty years after Barzagus first cleared the mountain, the mountain will appear as it always did. Only an archaeologist will find the pear stumps, old plastic drip hose, the charred remains of the shed, a tractor dragged over to the reservation, and the occasional rusted wrench among the second growth oaks and brush. It will be just about like the triangle of land near our tule pond, wild, untouched, and unwanted.

When I first saw the pear orchard only a few years after its creation—then, even in its abandonment, still stunning in its beauty—there remained unconquered massive stones and uncuttable monstrous trees in the field, which one was obliged to cultivate around, irrigate over, and drive beside when working the place. It was not wholly tamed even in the beginning. I imagine, then, from the huge stumps and piled stones, that the original layout was an untouched primeval oak forest. Even when these giants fell, their roots remained in the ground obstructing irrigation lines, polluting the orchard's soil with fungus, barring the path of pear trees, and resuscitating suckers from their stumps. Barzagus had purchased the land from a man, but in truth he had only rented it from the wild. What a strange notion it is to the modern farmer, that nature, its trees, animals, and waters, insidiously can still chomp on the bit, ever present, ever ready, all vigilant to reclaim again what it considers its own

In the valley below, that is all gone now or at least relegated to a ditch bank or dump site where the machine cannot enter, like our triangle of unwanted land. Our oaks, meadows, swamps, and herds have for a century now been cut, ploughed, drained, and slaughtered. And so the struggle in the valley has been transformed to a fight, not between nature and agriculture, but the even more desperate battle between agriculture and asphalt, between man and man.

A triumphant agricultural economics student once pointed out to me that an acre of houses—replete with six households' baths, showers, hot tubs, dishwashers, washing machines, swimming pools, shrubs, and lawn—takes less water per acre than a raisin vineyard, proving in his mind that there were no obstacles at all to the ultimate peopling of the valley. Why, he boasted, were not vineyards and orchards inefficient consumers of land and water when millions more of suburban men and women from Los Angeles might better take their place, when Fresno might sprawl and grow, acquiring a sophisticated, confident culture in the process? Might not vineyards and groves be uprooted, paved, curbed, sidewalked, and housed, to be renamed "The Vineyards," "The Orchards," or "Peach-Tree Hills"?

But even before the great drought, I had always thought that ecologically the valley could not support the teeming masses who in the 1970s and 1980s poured north into our agricultural preserve, polluting the air, taking the water, and paving over the soil, brains full of manicured nails, tans, mall strolls, sunglasses, and shorts. No, this university prodigy assured me, the farmers were the problem! Their valley for a century had actually been poorly utilized; its acres supporting a mere family per forty acres of vineyard and orchard when all along that small farm might have grown instead 240 families, replete with cul-de-sacs, Winnebago parking, and an occasional tree. For him, supposedly agriculture's apostle in the university, farming was but a child's diversion for ugly and inelegant men who had not grown up, not a billion-dollar industry that alone had never played California false during fire, riot, and earthquake, the refuge of the last sane men in a demented state of Menedezes and O. J. Simpsons. For him it meant little that the finest plums in the world, the after-dinner treat for millions of tired and bored Americans, were grown only in this valley. For him little more than jest was half the world's production of raisins, dried grapes that went into breads, cereals, and candy, that came from the great San Joaquin Valley of California and from nowhere else in America. For him that thousands of men and women pruned, welded, fabricated, and drove from the profits that accrued from such fruit were of no import if we all might yet become like those of his native Los Angeles.

I once met his master of darkness, Professor Bart Refino, the doyen of the university agricultural department. If pressed, Mr. Refino mutters a few words of rebuke as he darts on his way to the airport—Refino the international consultant, the local public's talking head about the mystique of farming, Refino the self-proclaimed world's expert on megafarm viticulture, at the airport, always on the plane. Thirty years spent advising corporations, of ridiculing agrarianism, of snubbing all but the progeny of local land barons had earned him an illustrious retirement, a gleaming endowed metal laboratory, an entire commemorative issue of *Vine Farmer* magazine, and sincere thanks from the chancellor for a job well done.

You know Refino. His species is interviewed each time the national media want to run a two-minute mini on the "loss of the family farm." Suave, now upbeat, he reassures you that the issue is merely "problematic," that loss is "too blanket a term," that there is nothing "romantic" about "agribusiness professionals"—the real saviors of the industry, who are indeed plugged into the information highway.

By late winter enough of the Barzagus mountainside was denuded to get half the trees into the ground the first year, with the idea that in the next February the remaining ten acres could be cleared and planted. The crunch then came in providing water before the young planted stock dried out by June. A local well driller, with special granite bits, was called in to tap the rock, hoping to find a hidden fissure beneath the stone that might contain enough water to provide ten or twenty gallons a minute. (A valley water well, in contrast, can produce anywhere from two hundred to two thousand gallons every sixty seconds, its volume for a while longer only limited by the size of the pump, not the availability of the underground supply.) After a $1,500 dry hole, the next drilling found a crack, and a small submersible pump went in (electricity had been strung in by the local power company weeks before). Even a tenth of what volume the farmer took for granted below was welcome, for many in the mountains drilled six, seven, and eight times without success in their search for water.

Like so many well-planned but doomed development programs in agriculture, the first year went well. As long as the struggle for farmers is strictly between themselves and nature—growing or not growing trees—they win. Defeat comes when the arena is widened, and the mercenary enters who must sell the new orchard's produce. When agriculture ceases and agribusiness begins, the predictable catastrophe ensues.

The eagerness and energy of Messingale meant that eventually the partners could rotate trips and thereby each avoid a daily commute. Messy's grafting skill ensured that cheaper (and hardier) wildroot could be planted and then, once thriving in the ground, grafted over to the preferred domesticated species of pear the next year. A local bulldozer was rented to carve out roads and grades, while the Barzagi (Bus and his son Buster) and Messy constructed a barbed-wire fence around the entire twenty acres to keep out animal marauders. In a matter of a few months, where once there had been only rocks, trees and scrub, open land for animals and vagrants, now through the wondrous practice of agriculture, over four thousand pear trees in neat rows appeared, the ground was tamed and ostensibly measured, then cut off by fencing from the surrounding environs.

To the untrained, nonagrarian eye, with aesthetics not food as its quest, the transformation of the mountain—like our own similar reconstitution of Knut Hendrickson's land—was startling. How could the junior and senior Barzagi, with the help only of Mr. Messingale, sculpt beauty out of rock, parched oak, and scrub? Terraces, alleyways, and bridges divided the various plots, as the pears—in the mountainous and nonpolluted air—achieved a deep green hue, unknown in the same species in the valley below. Had beauty, not fertility, paid dividends, the mountain would have been priceless, for it was every bit as silvery as a Cézanne or Renoir, this farm shimmering through stream and forest. Like our own Regal Red orchard, the pears' beauty was almost supernatural, as if it promised sterility as the price of its unmatched lushness.

It is not true, then, that man cannot improve upon nature in the narrow field of aesthetics. Should he understand nature, seek to master and to employ nature in agriculture, then the hybrid can surpass nature as man's plants highlight and contrast the wild about. Barzagus's wild oaks and dry maqui framed luxuriant pear foliage, as fresh red clay silhouetted dry brown grass. Lucretius, who wrote of the similar reclamation and domestication of the Italian mountainside, of the men that went up into the feral mountains and brought the civilizing orchards down, would have approved. But then he wrote poetry, not prose, not fiction, not history. And he was an Epicurean, not a Stoic.

Planting, clearing, and reclaiming, formidable tasks in themselves, were not, however, the chief problems. The land had to be farmed, had to produce fruit, had to create capital to pay for the expenses of its creation and serve the needs of its creators. The mountain pioneer can be a good land clearer, can stack his cut logs and pried granite, but he also at some point has to farm and do business. Problems—the age-old challenges of agriculture, like water, pests, and markets—arose immediately, suggesting that both would be difficult on the mountain. In retrospect, what sane mind believed that these young saplings could find moisture and nutrients in the granite soil? How could they survive the deer and the malicious hooligan, withstand bacterial, fungal, and insect attack, all to produce pears that men from below might pick, pack, and truck down to find their place on the table of the Los Angeles suburbanite? It was one thing to clear land and stick dormant sticks into the ground, quite another to make them live and bear fruit. Why would a mountain make allowance for something not its own? And even if the Pacific rattlesnake could be mastered as well as the mountain coddling moth, who could rein in a Buddy Messingale or Ross Lee Ford once the crop left the mountain? He might spray Guthion for *Coleoptra* or shoot *Ophidia*, but what, pray God, could he use on *Mercator Americanus*?

I can begin the array of disasters with the irrigation system. Miles of hose, plastic feeder lines, and main laterals were installed underground. Digging uphill in granite is not the same as trenching in soft valley loam. To push the water up and down the mountain through the half-inch hose, water pressure had to be maintained and homogenized through pressure regulators and time clocks. That meant the tiny pump would have to run all day, all night, automatically directing water through hoses among the various pear blocks up and down the slope. Barzagi and Messy dug, picked, trenched, and laid plastic for weeks on end to ensure that enough water was at each tree before the heat came in July. The pump might be a wonderful April and May pump, but could the same be said in July and August?

Remember, the trick in irrigation is not just the supply, not simply having sufficient water at the source. The rub is powering the water to *every* tree *every* time, up and down the row without exception. Irrigated agriculture is unforgiving, with a memory that cannot be deceived, must less cheated. Once the heat starts, the water must be there for the roots. You often overwater in the spring, knowing that by summer, when you can't catch up with the heat, you've already put some water there in reserve. Miss an irrigation, and the entire season is a desperate race to get even, your crop, even the life of your trees themselves, in the balance. During the drought, when there was no canal water, I once asked an almond farmer of two-hundred sandy acres out west of Caruthers how often he turned on his pumps. "They've been on since April," he told me in August.

On the mountain, squirrels and gophers sought relief from the dripping water and so ate the hoses, occasionally ruining the pressure in the system and sending precious water down the mountainside. The homemade filter often clogged from the granite flakes brought up with the water out of the rocky ground. More worrisome was the incapacity of the supply itself to achieve the needed pressure. Water from the pump could keep the hoses full and oozed out a gallon or so per tree every six or seven hours. That might keep the trees alive, but was it enough to produce and size fruit of the quality of valley pears below?

The answer clearly was no, especially as the seven-year drought wore on and there was no winter snow to replenish the fissure. The trees then might not die (ironically, without a crop they might look better), but like a tired nursing bitch, they would either shed or not size their own offspring. The hunt was on in the mountains for a supplementary water supply, the original well already tapped, exploited, and found wanting. Without fifteen gallons per tree per day, the pears would be closer to pear-plums than pear-apples, undersized, bitter fruit that could not be consumed and would not be wanted by either Buddy Messingale or Ross Lee Ford.

Whether because of water shortages, frost, or idiopathic failure, the first year's crop failed to set, even its few surviving pears not making size. The concurrent raisin crash below heightened the crisis. It's one thing to be diversifying when you may be breaking even, quite another to be failing at both your base and your extension, both your raisins and your pears. Still Barzagus, the fig man, the peddler, the raisin stemmer, and the contractor, answered each pear catastrophe with extraordinary ingenuity and renewed exertion. Buster his son devised new pumps, water lines, filters, and timers. Messy investigated the esoteric arts of fruit pollination and maturation to understand why trees that each set one thousand five hundred pears in the valley below would not hold fifty above. Bus would find the water.

He dug another feeder line, with accompanying electric cable, nearly an eighth of a mile through the granite, across the property line, under trees and brush, onto the neighbor's creek. He would pay the absentee owner fifty dollars a month for his surface water and as an added gratuity build an automatic watering trough for area cattle. (But why avail—even nourish—grazing cattle that might break down fences for a sweet pear?)

The exploitation of the creek also posed questions of conscience for an itinerant lesbian college professor and her cortege of ecological student explorers. The girls all went up to wander around on retreat and to ride beside their mentor each weekend. (The local university's stern code barring sexual liaisons between students and professors is exclusively aimed at the gutted and balding, the fifty-year-old slob who drools, cajoles, and occasionally paws the scantily dressed coed; it has no teeth over the more elegant and sophisticated homophile.) At the very outset, the girls had naturally felt outraged by Barzagus's intrusive male presence and the original masculine scarring of the folds of their mountain paradise, the crashing down of ancestral oaks and the splintering of picturesque boulders. Worse yet, the idea that the white male once again would cause discomfort to the Indian was fodder for a university sensitivity session. And the latest creek siphoning in particular disturbed them to no end (his pump was now right near their only quiet and secluded creek beach) and called for special amends where necessary. So Barzagus devised an easily accessible, constant water source for their horses (a flotation device hooked to a trough supplied by his irrigation system) and proffered his former academic credentials as proof of his environmentalism. By year's end, when the aged professor died of breast cancer and her college ridership dissolved, Barzagus was probably considered a kindred mountain recluse rather than a nineteenth-century destroyer of Sierra earth, water, and wood. I suppose too that as a former academic himself, Bus knew that the majority in the university can be bought off for ten cents on the dollar.

Unconquered, then, by man and beast, he at last dropped a bigger submersible pump right into the flowing rivulet and pumped the water back up to his mountain to supplement his own failing underground supply. The two pumps together might give a fifth instead of a tenth of a valley well. Still, even that effort seemed not enough; the next season only some fruit set and they did not size. There was now no more water either above or below the ground.

At this point Randy Messingale wanted out. The thrill and diversion inherent in pioneering was now over. The mountain wilderness wondrously devoid of pedestrian farmers turned out to be replete with Indians, Okies, cattle, and strange horsebackers. If he was not to be a Sierra Nevada mountain man, it was better to be a nondescript farmer near the river where costs could be known and mastered. The gamble at a quick success had surely failed. It was now a much more complex, more expensive question of producing profitable fruit in the midst of a growing depression (it was now 1985, two years into the raisin crash) under embarrassing circumstances, miles from home.

Messy was not making money below. So the hour's drive in order to lose more money made him calculate very carefully, family connection or not. Mounting expenses, like more hose, plastic pipe, pumps, power, emitters, replacement trees, diesel fuel, stolen equipment, blown tires, and missing batteries, made him especially nervous. If Barzagus and he both sold (or more likely, simply walked away), they both lost their tiny investment. If one stayed, could not one get out? An expert in arboriculture, he foresaw that there were more fundamental, unsolvable problems in these mountain pears other than water—like fruit setting, fire blight, and hard frosts. I often thought later that he must have known, too, that the orchard had now no intrinsic value at all either as a productive agricultural enterprise or as a vacation getaway for the elite.

Messingale, the near relative but always the hard nosed, was in a dilemma. Open and candid admission of the depth of their joint error and lengthy discussion could result in sudden abandonment by all of the project and forfeiture of his own small investment of time and capital.

On the other hand, silence, melancholy, and the growing excuse of a new wife and baby might camouflage his suspicion of failure and allow his partner to buy him out and press on solo. On a rare occasion, in order to get out himself, might not Messy even feign an interest in cashing out Barzagus? If done quickly and halfheartedly, the idea of buy out at least introduced the principle that the land was too small for two and had earned real equity. Then his own preordained exit, if packaged as a favor to Barzagus, might save his meager invested cash. (In Messy's way of thinking, in a depression even $15,000 can mean the difference between collapse and revival. And so it often can.)

Messy came up less regularly, and when he did, he complained incessantly about the mountain. Soon, all too soon, Barzagus reacted grudgingly to Messingale's (clumsy enough) efforts, paying him with interest for all his labor and capital, eliminating a whiny critic to tell himself what he already suspected. Better to ruin the health of one than two; better to pay than not to be blamed at family gatherings; better simply not to have Randy Messingale on the mountain at all. The fruit was not setting, and what did would not size.

When I saw Messingale four years later, he was smug that he got out of the pears without financial loss, as if his absence on the mountain was proof of his own agricultural foresight, not the result of misguided Barzagus charity. He still crowed that he would have liked to have bought out Bus. He could do that because despite the farming crash, despite the unfortunate end of Buddy, life had not been cruel to Randy Messingale in the postmountain years. His new wife had some family money. But far more importantly, the city had now moved out to his ranch perched above the river. Not the poor or minority city had met him at his water, but only the richest and whitest who might stay in California after all, if they could build bluff estates on Messy's hill overlooking the cool running eddies. Messy might soon be devising not orchards and vineyards any longer but lots and executive homesites, with the chance for a million or two in cash, more money than his entire line had made in a near century of farming. Could not a mountain man who cut oaks and dragged out boulders just as easily survey lots and engage developers? Farming houses was far more lucrative than producing food. Why grow plums and grapes for the Wilsons, Garcias, and Yees when you can bring them onto your place and plant them instead—house, car, and kids—right where your vineyard and orchard used to be.

I concede now in abstract terms the Barzagan twenty-acre fiasco was very small beer compared to the collapse of thousand-acre empires and unpaid million-dollar notes in the aftermath of 1983. But in reality, corporate reorganizations and recombinations have little drama outside the business pages of the local papers and are of no didactic value at all in understanding the accelerating transformation in American farming. They are mere Rubic cubes for the local ag econ department. Farm directors are fired and resurface with different complimentary pens and calendars. Losses are charged to distant investors and deducted from government taxes; dollars budgeted back east to California agricultural investment by insurance conglomerates shrink and swell. There is never any war there between the landowner who works the soil and the land that is or is not mastered. Never is it much different if the land is near Fresno or south of Bakersfield, beside Five Points in the Westlands water district or right outside of Coalinga. Wherever or whatever the large tract is, it is not farming, a profession whose primary purpose, according to Xenophon, was to serve as "the best tester of good and bad men." The struggle over federal water of the R. L. Rockwell thirty-thousand-acre corporation tells us nothing about farming; it can only fade before the more epic fight of a cursed Phong Her on his pathetic ten-acre cherry tomato farm. Seen under that light, the mountain tragedy of Bus Barzagus was every bit as important as the liquidation of a MacCalister Farms Incorporated, whose monumental collapse at the time dominated the agricultural pages of the local papers.

Birds began eating the few small pears that did appear, voracious Stymphalian-like mountain birds, mutant winged species who defecated constantly on the scant fruit and made bizarre noises. Mountain crested jays, woodpeckers, blackbirds, crows, ravens, and starlings welcomed the open orchard, a tasty relief from the grim, bleak, and cold surrounding forest. Barzagus researched cannons, sprays, and scarecrows and in the end draped all the pears in acres of plastic netting. (My suggestion to purchase one of our surplus Effin balloons cum plastic chicken-hawk kites was politely declined.) From then on, he did all his cultivation and harvesting of the trees under wraps.

But was there really much to protect? The fruit was scarce and undersized. I consoled him with the dictum (absolutely true but which I myself could not follow) that in a depression it was better not to have than to have fruit. At least, there is no harvest cost to lose, no roll of the dice to suck you further in, no Ross Lee Ford to blame.

Ralph Wilson, a little man who owned the tire store in Alma and played around on twenty acres of plums, hadn't seen it that way when I, the veteran of lost causes and lost equity, pointed out to him the economics of depression farming. As his man fitted on an overpriced tire, he bragged to me that he had produced fifteen thousand boxes of Santa Rosa plums and would have made money if the price had been the old twelve to fourteen dollars rather than four dollars a lug. He beamed that he'd have twenty thousand boxes next year. "You can lose even more," I offered as a compliment. "I'd rather have one tree in the yard and lose ten dollars than have your orchard of two thousand seven hundred." Who but little Ralph Wilson wants to feel big when that big breaks you?

Deer, cattle, and an occasional bear browsed fruit and broke limbs more boldly in the absence of man. A local delinquent was hired to take an occasional shot and turn on the pump. The barbed wire was patched, but again the problem was with the fruit itself. The trees did not want to set, and what fruit did appear did not size. Even if a big crop had set and had sized, Buddy Messingale, the inside broker, the L.A. fruit-market hound, was now holed up in a mountain trailer, cocaine-ridden in the slow but inevitable slide to cardiac arrest, a few months before his demise constantly peddling fantasies on the phone to no one in particular. If he ever did get hold of your pears, they might end up as collateral for some drug deal in Compton. You'd be better off giving what little you had to the local consignment salesman in Alma for two or three dollars a box. Ross Lee Ford—who Barzagus reasoned was less preferable than Buddy Messingale—sold our nectarines and persimmons once for little more than a dollar a lug, recovering for us the price of the cardboard box, losing the three-dollar-a-box harvest cost and the four-dollar cultivation investment. Still, he sold the fruit I suppose.

Ugly beetles and worms worked on the trees and drilled holes into the fruit. Old-fashioned organophosphates were called in to protect a crop that was mostly not there. It was best to get the most lethal brand. There was no need for integrated pest management up here, no worry that valuable predators might perish in the spray. These evil-looking bugs would require the strongest dose necessary, perhaps a shot of Lannate, Diazinon, Defend, or Dibrom gas. Better they all die with one whiff.

Lethal rattlesnakes were predictably hiding in the shade of the orchard, sipping the scarce water as it oozed out the hose. In response, workers turned hunters and smashed them with rocks as they searched in vain for fruit. On my rare visits everyone walked with eyes down.

Frosts were found to be far more severe in the mountains and decimated the April bloom. Plastic polymer sprays were used to coat and insulate the trees while the tandem pumps were kept going constantly to bring the warmer water out of the ground to create a little steam under the trees. Fruit might now more often set, but it still would not size. Barzagus noticed that even if the frost did not kill the young pears, it shocked them and retarded their growth.

Vandalism continued, but now the more successful reverse strategy of enticement was used. Everything on the mountain was left wide open (now for the most part of no value) so that the locals could see the shed was not padlocked but empty, the tractor still in the field but without a battery, the spray rig accessible but so antiquated as to be of little value. Yet was that any way to farm, to appeal to the calculating side of a thief, that his dramatic effort with the bar, hammer, and match was simply not worth the meager return? Still, the native Sierra young men left most of the fruit alone because it was under nets, too hard to find and of small size, not at all up to the standards of the discriminating gleaner.

The mud and roadwork were remedied by the neighbor with a bulldozer. An old tractor was sold cheaply for his work at regular maintenance. Anything that went up to that mountain and could not be placed in the back of a pickup never returned to the valley below. The road in the late fall was now manageable, but there was no reason to go in when there were no harvests. The mud was not like valley loam when it gets wet. It was red and sticky, mixed with pebble and rock, like concrete. It swallowed axles along with the tires.

The real, the only problems of producing marketable fruit were never so easily solved. Fire blight, despite copper dust and antibiotics, gradually grew endemic, not better. Whole blocks of pear trees now blackened and were bulldozed under by the neighbor. Damaging frosts overcame sprays and water and limited the crop. The few pears that did set still did not size, as both pumps, creek and granite fissure, lost, not gained the necessary volume. Barzagus simply turned one pump off once the trees started dying. The drive up the mountain grew longer as the quest changed from growing pears to preserving a few trees that in the distant future might grow pears.

A mere three years after planting, the pear market itself, like the raisin, pear, plum, almond, walnut, grape, peach, and nectarine market, collapsed. Boxes of top-quality, choice variety, enormously sized valley fruit that twenty-four months earlier went for thirty dollars were now selling for six and four. Oh, they were still nearly a dollar per pear in the store (now the equivalent of $48 a box), but the wholesalers soon had enough of a supply to pay whatever they wished, and the psychological tide shifted to the buyer. Barzagus's lesser mountain brand, even though they came off late when the market was cleared, would not even meet the cost of picking and trucking. He contracted with a peddler to sell what little appeared on the trees at farmer's markets.

Without a crop, he went up to the mountain less often. The last time I drove up with Barzagus, an enormous Galapagos-like lizard, a horrible looking, foot-long dinosaur, stuck his bony crested head out of the spray rig, where he apparently had been nesting in the bottom of the impotent chemical residues. Of all the hazards on the mountain, who might have envisioned the possibility of a crushed, shredded, pesticide-laced, cold-blooded carcass clogging the sprayer pump and nozzles? I pointed this incongruity out but was admonished to look down, not out, because a family of rattlers had recently moved into the lower orchard.

Now in the midst of the farm depression, even if the deer, the cattle, and the bears were scared off, the minutemen calmed, the Native American youth pacified, the water pumped, the bugs poisoned, the tractor charged, the girls and their horses won over, the mountain commuted, the road cleared, the gophers, raccoons, and squirrels killed, the birds netted, the men driven up the mountain to work, the snakes stoned, the frost warmed, the fire blight stopped, and the fruit set and sized, no one at all cared to pay for the labors of that harvest.

The farm crash meant that we were already losing money on our own beautiful apple-pear orchard a few steps from my brother's door, even though it set and sized beautiful fruit without the snakes or the vandals.

Barzagus still for a few years kept a tenacious hold on his amorphous mountain creation, despite his money-losing raisin ranch below, despite his newly rented fig orchard, despite his fruit-peddling on the coast, despite his construction project in Fresno, despite his raisin stemming and processing machine. While the orchard breathed, there was life. While there was life, there was not yet failure, as the farmer, every American farmer, like a beleaguered general down to his last division, now grasps on to hope as long as he can, to fool himself that he has not really lost money for the privilege of sacrificing hours of his stoop labor.

As long as the loan is met and the interest paid, all losses are still only theoretical, on paper alone, mere dark private thoughts in the farmer's brain. That delusion only becomes a dreaded reality, the nightmare only materializes and takes tooth, when the farmer awakens from his coma, takes his medicine, bellies up with the loan officer, and settles his crop year each fall with the bank. Men with new pickups, new houses, new tractors in their vineyards, tough men with sour faces and the steely one finger raised from the steering wheel as their sole greeting, they can all crack and shatter in the fall.

October is the cruelest month, I think. One day they are there, the next they are not. Mere shadows, they flit down to the agrarian Hades to be reprocessed and sent back up, transmigrated into the nonfarming majority. In a matter of hours at the lender's docket, the yeoman running on low equity finds himself reduced to serfdom, peasantry, or fixing tires in town.

That is why, I think, many raisin growers in the 1980s did not come into the bank after harvest and why their remnant a decade later still sometimes hesitates. When the short crop of fall grapes are still drying on the ground, the doomed know instantaneously that they are to be targeted and know how they are to be found out. They school together in the cracks and eddies of the bank's clumsy postharvest shakedown and size-up. To harvest the errant agrarian, the computer men in town have to snag and reel the horny handed in fighting, with threatening letters, phone calls, and summons if need be. The foreclosure or loan termination letter, *crede mihi*, is a frightening weapon in the banker's arsenal. Thirty days and everything that is yours is theirs. Vocabulary appears that not even a linguist can fathom, much less those who sent you the letter in the first place. You do notice for the first time that the fall prior, you signed your name on twelve pages twenty times, signed away your house, land, car, and salary—signed away your life.

Still, the failed fight the current. They are only netted and beached when the money runs out, the crop advances cease, and the unpaid men and machines grind to a halt in their vineyard. As long as the farmer doesn't walk into the darkness through that awful tinted-glass bank door, he is not yet found out, not yet gone over to the majority. Might not raisins still go up in price during October to December, he reasons? Might not a Halloween fire yet abolish the bank's records? Might not by Thanksgiving, I once thought, a new, gentler loan officer appear from out of state to roll over the debt, to look at the farmer's hands and say, "We'll work with men like you"? Might not the raisin packer's report be lost in the mail on its way to the bank? Might not the bank's figures themselves pencil out differently from his own? Finally, might not by December 23 someone with suit and tie say enough is enough, we will do this squeezing no longer, might he not agree in the end with old Petronius that "we are men, not gods"?

At least Barzagus drove up for a while longer, if only to walk around and stare for ten maybe twelve minutes and then leave. More and more the deer browsed, the checks were mailed out to Messy, the pump clogged, the hoses separated, the birds found holes in the net, the underprivileged destroyed, the fire blight burned, the weeds grew, the insects ate, the snakes drank, and the fruit either did not set or did not size as the mountain gained confidence and beat the farmer back. As it had grown acre by acre, so the tiny pear ranch now shrunk by twos and threes. In the end a single block endured, hogging all the well's water, incongruous in the midst of the stumps and black skeletons of its dead brethren, the reminders of its own fate to come.

At last a change came over even Bus Barzagus, agrarian zealot, self-made farmer, undefeated veteran of twenty raisin campaigns, keen student of economic cycles and human nature. His hearing became bad; a chronic cough developed in his lungs; his ankles were swollen; and he was not yet fifty. Cigarettes of a three-decade habit brought not their accustomed buzz but nearly sickened him. He sought relief in sulfa drugs and sunflower seeds. Where before his eyes were crazed with some new device to battle all those who would destroy his pears, they now were empty and dull. Where before he cursed the raisin industry, the banks, the corporations, and their diseconomies of scale, he now said the battle was over, defeat assured, and an armistice necessary. Where before he talked of an eventual return to inflation, he now groused about decades of worse and worse deflation. Where before he would shudder in rage at Safeway's $1.89 a pound peaches (twenty cents going to the farmer for his expenses), he now chuckled in despair. And where we had once sought solace for our own far greater blunders and missteps, we now found only greater confirmation of our own futility.

A terrifying scowl met any who offered hope, as his entire life in farming now was reduced to a fool's delusion: survival on the land was to require meanness not solemnity, miserliness not frugality; optimism in farming was revealed to be derangement, faith imbecility. The curtain had parted, and Mr. Barzagus had glimpsed the preordained Armageddon of the American yeoman. The view was there for any brave enough to step forward and take a peek, but most preferred the secondhand shadows on the wall to the searing torsos in the lake of fire. For, as Mr. Barzagus knew, the answer to the destruction of agrarianism lay deep in the bowels of a larger American sickness and involved the Golden Calf we have worshiped for two centuries. To recognize, to acknowledge that malady, that America made no connection between food and farming, required a near denunciation of everything one held dear. And that was a truth few farmers other than a Barzagus had any belly to embrace.

When the primordial urge to work the soil finally abandons its rare agrarian custodian, beware! That spirit flees and leaves the limbs to sag. Even the most devoted, the most ingenious, the most successful farmer can be hemorrhaged dry, left a zombie after five or ten years of continual tapping and siphoning, by the realization that he is failing as he is borrowing scarce money—stealing really his wife's paycheck, his dad's social security, his daughter's baby-sitting money—for his now selfish struggle in the field. When this realization arrives that the yeoman is a leech, a barnacle, a self-absorbed mollusk, it is like a divine bolt to the cerebellum, the voltage convulsing the body and clearing the slate clean. From that point on, the agrarian, if he survives the shock, can never really farm again, for his noble break-even struggle has now turned traitor, become an ignoble mechanism to rob everyone he holds dear. Oh, he can perhaps irrigate and even drive a tractor, but it is all mindless puttering without purpose, done without pride or skill. The urge to do some great thing, to conquer the elements in some heroic fashion—vital to even the most pedestrian of agricultural tasks—is extinguished, and when departed, even the mundane and the everyday processes leave as well.

I think even the most reactionary and Republican finally sees the elephant once he knows that his four-dollar plums sell for thirty in the store, once he understands that men of his own kind at the bank are to give him thirty days to clear out, once he understands that his land shrinks as Ross Lee Ford's latifundia expand. The blank stare even takes hold of a Carl Warkentine; fourth-generation Mennonite farmer with John Deere tractors, blacktopped packing yard, two hundred acres of white ash soil, the finest, most pretentious farm in the valley; he at last shrugged even to Bus Barzagus that "he made a little here, and then lost that little over there." The most conservative farmers in America—men who had wanted so dearly for so long for that decade to appear—not the organic tinkerers, not the New Age investors, not the owners of boutique vineyards and apple juice farms, were obliterated by the farm shakedown of the 1980s, by the new gospel that food was to be quite lucrative for everyone except those who themselves were to grow it. The eighties were to conservatives what the sixties had been to liberals. Be careful, the Greeks warned, to wish for what you should not have.

When Barzagus finally came down off the mountain, abandoned the orchard, turned off the pumps, put a realtor onto the land, paid Messy off, his body then—but only then—collapsed, and in the last hours of his health he was wheeled in to be saved and rebuilt at Saint Agnes Hospital in Fresno. He survived the heart operation. After a year, with far less cash and health, he turned over to his son Buster his money-losing raisin ranch two miles to the south of my house, and he did not go up to the mountain any longer.

We two defeated, we two nearly nonfarming now, talked just today on my porch about all the things we once did to save us that have since all but destroyed us.

CHAPTER EIGHT

THE AGRARIAN PANTHEON

These last and final generations of family farmers have not been saints. Most agrarians are obtuse and blunt, not sensitive people at all. Much less are they well-read students of history, hardly at all inquisitive liberal minds with a need to hunt down the truth, ignorant entirely that they are now targeted in the crosshairs of the complacent American public. Need I add that few are physically attractive persons, not clever in speech, not imaginative in dress? Theophrastus twenty-four centuries ago rightly saw in the agrarian the archetypal boor, oaf, and clod, who "will sit down with his cloak hitched above his knees, exposing his private parts. He is neither surprised nor frightened by anything he sees on the street, but let him catch sight of an ox or a donkey or a billy goat, and he will stand and gaze at it."

Often, I think, yeomen lord their freedom over you, as if everyone gladly chooses to stamp out hubcaps or all day to play phone tag with their bosses' snooty clients, as if the choice to work the land is free to all who are born in cramped apartments or who follow their dad from duplex to the plant. Farmers, as you have learned, complain about the injustice of their plight, the robbery of their produce, the cruelty of nature, and the carnality of the bank. Peel off those justified layers of reproach, and beneath there is at worst a smugness, at best a serenity that comes from having no master, and the exultation, even ecstasy, of placing their all, muscle, brain, and nerve, on the table against nature. Even the odds that say it is more likely now that you will fail and not pay back the bank can lend an electricity, a mission, an urgency to your day, as if you are marching out to face the legions of doom for the worthy cause of feeding your brethren.

That allurement of the land and the confidence given those who would work it is why the land does not go out of production when families fail. Other queer urbanites of dream and fantasy stand ready, tyros all, preened and lined up to take their place, oblivious of economic reality, eager to pledge their savings for a while to become agrarian, to try a hand in Aristophanes' rural world that is "paradise, untidy, easygoing, and unrestrained." Even the corporation and hobbyhorse farmers cannot ingest all the debris of failed agrarians. Even the perpetual renter and the vulture who farms the repossessed land of the bank have had their appetite sated, even the influx of Punjabi vineyard investors cannot handle the acreage on the market.

You see, there are four or five million Americans—and foreigners as well—out there, with money, with capital, who envy the life of the farmer. When the home place goes up for sale, they can be lured in to try their hand at the roulette wheel, to ensure that more food continues to be produced at a loss. When they are exhausted (in a mere five years rather than the fifty of the yeoman), yet another cohort sits waiting. For there are less than a million family farmers left in America, and three hundred million of the other. Of that other, a percent and a half of the population would like to work for no master and imagine the wind in their face as they produce food. Between the corporation, the weekend farmer-teacher, the hustling renter, and the eager novitiate, the land gets farmed. No wonder the agricultural economics department says not to worry as the agrarian slides away, for are there not "too many farmers," is there not "excess production capacity"? In some sense, there are.

Forget that the yeomen are the most efficient farmers in the world, the men and women who alone retain the American mark of excellence in the shark tank of world trade. As long as others stand ready to farm with nonfarm capital to lose, the market forces that balance land with farmers will continue to work, and ground will be worked at a loss. I thought Bus Barzagus was somewhat obscure, though correct, when he noted that family farmers plough all the vineyard's income back into the land and take little of it for themselves, while their various replacement species take most of the grape's output for themselves and invest only the capital of others, the nonfarmers, into the ground.

But you more urbane readers, I imagine that you would not want any of these family farming holdouts at a university lecture, a golf outing, or a group therapy session, much less on a weekend retreat or conference panel. Perhaps they would be embarrassments even at chili cook-offs and neighborhood watch parties among the carpenters and truck drivers. You see, as Theophrastus knew, the tanned crack in their behinds too often peeps out from their sinking Levis. By forty a key incisor or two is left broken and turned a lifeless, a very ugly purple. Their drooping gut pops the last shirt button below the navel and at times is left partially exposed, sliding down, hiding the groin itself. Their winter sleeve is crusted with mucus or worse, for they are around trees and vines, welders and tractors, and few folk enter their cosmos outside of town.

They track mud on your linoleum and leave dirt on your sofa. By them the young, sensitive homosexual who grew up across the road is dubbed "the boy who went queer up in the Bay Area"; the wealthy Armenian raisin packing elite are reduced to "the long-nosed thieves"; the grape picker on permanent disability with the fused back is greeted with "always siesta, huh, amigo?"; the industrious but undercapitalized farmer without inheritance is cruelly dismissed as "the know-nothing about to go belly up."

Indeed, I often find the agrarian downright uncouth and unkind. All that one can do in his limited station off a farm—teach school, write books, counsel the young and timid, tend the sick, or design parks—is reduced by the yeoman to "city work" of little consequence and value if his hands have gone smooth and lost their horns, if his health is shot or his belly for rising at dawn full. I have seen a state appeals court judge, a graduate of Stanford University and arbiter of capital cases, renowned jurist in her sixties of some repute in Fresno, berated for a missed date at the farmer's market, for not playing Euripides' mother barking in the agora, for seeking a reprieve from peddling the family fruit at the local produce bazaar for one hundredth of her normal hourly wage. Agrarians are not perhaps spoiled, probably not even arrogant. They are, however, unkind to their own, especially their benefactors, who are to sully themselves in an evil world to save their land.

Instead, the incontestable nobility of vanishing California raisin and fruit farmers, like the valley garter snake, the mountain fire newt, and the foothill kangaroo rat, lies mostly in their endangered status. They are walking anachronisms, more ancient and natural folk out of kilter with everyone they meet or see, soon to be obliterated by an urban world they do not fully understand, a twenty-first-century world that has no place any longer for an embarrassing species of their ilk who know little more than when and how to water vines, to patch vineyard wire, to thin plums, to produce fruits of all species in limitless volume.

This random agrarian survivor is never named Ashley, Boston, or Justin. More likely for these men in their fifties and sixties try Harlin, Pete, or Calvin, his wife a Hazel, Agnes, or Doris. Even the rarer younger yeoman does not curl or dye his hair and bears no artificial scent other than the perennial musk of diesel fuel. His muscles are not sculpted by machines but hidden in his back and ample belly. It is not untrue that he eats messy ribs; marbled steak is preferable to the tossed salad. By fifty his teeth are not capped or even filled but can be jagged or pulled for ugly plates. There are never stones nor gold in his ear, around his neck, on his wrist; even his fingers are bare of metal. I imagine his underwear is not tight and red but boxers white and baggy. His truck has no bed liner. It is not jacked up a foot or two in the air with ostentatious roll bar and useless fog lights. It is filthy, dusty, dented, and littered with coffee mugs, cigarette butts, twine, shovels, scrap iron, and used batteries.

The yeoman says that the nematoded vineyard is "played out," his sputtering tractor "shot." When he overfertilizes and underwaters, he has "screwed up." His short crop is clear proof of "failure," that he is "no good." A money-losing year is evidence that someone else, someone "smarter," might take the land and do "better." In the world outside, the shiftless are "lazy," the less bright "stupid," the dishonest "crooked." The alcoholic is a "drunk," the drug addict "better off dead," the lame unfortunately "crippled." The abused child who grows up to rape and murder? He has earned, not thousands in subsidized legal briefs and psychic evaluations, but a long overdue date with the hot seat or a mouthful of fuming pellets.

When you fight nature, natural not social science becomes your yardstick. You do not "come to terms" with roller worms; you are not "uncomfortable with" a killing March frost; you do not "feel the pain" of a desperate but water-thieving neighbor during a drought. You learn about an unforgiving humankind from observing the ways of nature, the growth and decay of your own orchard and vineyard. Let the other 99 percent think that they can learn of real man from observing human not agricultural law.

And so of all species of American, most unlikely so, the farmer, not the university-educated—much less the counterfeit and pedantic classicist—turns out to be the true Greek, the sole kindred to Plato's forms and Thucydides' chance, fate, and necessity, the heir to John's voice in the wilderness. For the unhyphenated man quite apart from others believes in bitter but divine absolutes, in Antigone's unchangeable laws of evil and good, in the need for culture to fight the good fight against nature, the soil man who sloshes defiantly through the stagnant pools of -ologies, -zations, and -isms, oblivious to, unconcerned with, untouched by the ooze of pretense, cant, and fad.

When pressed and cornered, the agrarian does not give the thug's death stare nor the rich man's look of dismissal and fright but merely bewildered anger. When I see a rare *Agricola americanus verus*, an

Agrarian American, wandering in Fresno far from his den, I am both amazed at his excursion and shudder at the same time for his unabashed bluntness, as if people in town can still be like trees and vines, to be dueled with on even ground.

"You betcha, I'm hungry," I hear as he answers the curt waitress. To the doctor with the lethal diagnosis of malignancy, he shrugs, "Well then, Doc, I guess I've had it." You turn in surprise as he alone confronts the crowder in the DMV line, "Play fair with me, mister, and I'll do the same with you." "Well, then, son," he answers the aggressive foreclosure officer at the bank, "go ahead and take the place, see if you can farm it better yourself."

He is no different from the rare bewildered and bedraggled kit fox that appears mysteriously for a second in the vineyard row, stunned and unyielding before the nighttime tractor's lights. He is now alone and lost, his territory, his haunt moribund, this family farmer on the eastern side of the urbanized San Joaquin Valley of California. Like the endangered valley snake, newt, rat, and fox, his demise is nevertheless oddly majestic, for like these pedestrian animals he does not change a regimen that is to kill him. Nor will he whine or blame as he loses his land. Though he be broke, defeated, and bankrupt, give him yet one more year, one last line of credit, and he is ready nonsensically to produce food in money-losing surplus once more.

We, who have no need of this unpleasant reminder of what we are not, do not give him another year. And so we dub him—who in the actual production of food is the most efficient cultivator in the world—inefficient, outmoded, a loser of other people's money. We eat his plums for the price our parents paid in 1950. We snack on raisins for which he gets less than he did three decades earlier.

We subsidize instead the water, the land, and the produce of the absentee corporations even as we damn them, for are they not better paid, better scrubbed, and more agreeable—like us? For ice tea on the lawn, few of you would prefer the forty-acre man, gummed Hank Sutherland, bibbed overalls, master lecturer on the chance of a southern storm this harvest, hearing shot, back shot, knees shot, to crisp Rick Heavner, trim manager of Stonegate Vineyards, Incorporated, Rick Heavner degreed, bejeweled, immaculate, pompadour hair, smug with keen nonfarming interests such as hang gliding, golf, and rock climbing, jet skis, not a hay rake and rusted plough, behind his house.

But even within that vanishing breed of agrarian like Hank Sutherland, there is also a more peculiar handful of valley farming people, who rise above their own doomed kind and so tower even farther above the rest of our citizenry, who alone belong in the agrarian pantheon. Often I think a chance aberration must have set them apart from the lesser American agrarian *populus*—sometimes a darker hue, occasionally an accented English, more often simply a physical deformity such as a stutter, shrunken hand, or bothersome squint.

Anything that might draw the odium of the pack is the key that ignites the reservoir of untapped power, the catalyst that gives these lions chance for isolation and reflection, pause to see that they are in fact *not* like the rest of us in other—far better— ways as well. And so these are the rarer flakes of precious metal that have washed out from the American agrarian landscape, quieter men and women who have put together all the good things that yeomanry instills and filtered out the coarse to fashion a self-deprecating but unquestioned courage far superior to the more blustering, more formidable farmers in their midst.

They are not consensus builders at all, so completely immune are these queer agrarians from anything resembling fashion. They are not brainstormers who make lists of pros and cons to impress supervisors and overseers. They are hardly nuancers who worry over the shades of morality.

No, they are often crazy, sometimes dangerous Manicheans, who know that like a farm's growth and decay, fecundity and sterility, labor and sloth, productive and unproductive vineyards, there is only good and only evil in this world. The in-between is for others, a now-fashionable mirage of the corporation, government, and university especially, a chimera that arises from contemporary ignorance or from the timidity of the nonevil who have lately become too weak to do the good that they know to be good. Timidity is, you know, the price of our nonagrarianism, the inevitable tab when millions of people exchange their hours of monotony for electronic playthings, claim silly titles as their due, judge their worth by the location and size of their house, and believe there are such things as raisin plants.

These other agriculturists who still prefer to produce food at a loss have been singed, scorched by the agricultural inferno, but not yet altogether blackened. It is entirely fitting to end on a note of eulogy for those few left in California family farming who have tasted the dust and yet have become cleaner, not more soiled, for the fall. They were not just better than the rest of us in agriculture; they were superior to anything our exhausted country, our fourth-century A.D. fragmenting empire, had left to offer, the vein that lay untapped with roots leading down to the very bedrock of the now-forgotten agrarian republic.

I. Taro Kobashi

I discuss now the living, not the dead, not the giants of the past like Rhys Burton and Hans Hendrickson, and so I start with Taro Kobashi, raisin field man for a recently bankrupt packer. I did not know him even by sight before we contracted to sell him our raisins after the reorganization of the cooperative. I scarcely know him even now. I have not seen him for a year. Nor did of any of us have much desire to talk, much less to socialize, with Kobashi once we signed his company's papers. Most farmers with any sense, as I have implied throughout, don't like field men, company reps, chemical salesmen, fruit brokers, and insurance people—even if they be family and kin. They legitimize agriculture but themselves never produce food. Those time wasters talk beside their cars, in their cars, on their cars, always talk when they might have pruned twenty vines or harrowed a few rows of vineyard.

For their part, these handlers and middlemen who dress like Taro Kobashi do not like to farm, at least to farm after prior failure. Privately, they usually disdain farmers who know the ways of plants but not their more important customs of business and markets. With some cause, they see us as dreamers and as childlike, we who grow up on the land oblivious to the harsh financial cauldron around, clinging to the agrarian womb that shields us from the eight-to-five nightmare at the edge of the property lines. "You farmers," a disgusting and disgusted academic once told me in disdain when I quit classics one afternoon in spring 1980, "you put a toe in the water outside your birthing place, don't like the chill, and then you run all the way home." But remember, Emerson did say that "all men keep the farm in reserve as an asylum where, in case of mischance, to hide their poverty—or a solitude, if they do not succeed in society." A seventy-five-year-old farmer down the road once beamed that one day he might still own his dad's place (who was ninety-four). To date he and his wife had bought neither house, land, car, nor anything much other than what was cast off by his parents. I wondered whether he had even been to Fresno.

It is ironic that the finessers' job requires them to stalk down agrarians, even to flatter them, worse yet to romanticize how they might still trade places. If the salaried lackey, usually now named Joshua or Jason, visits a farm on an especially busy day, when a semi is in the yard loading twenty tons of raisins while ten men are packing persimmons, while late Sierra water is running and the big ditch gates are open one last time, while children are palletizing and banding lugs of pomegranates, while the tire man is fixing a huge tractor flat and someone is welding a broken fork on the lift, then they become addicted to the frenzy. Indeed, they become downright polite, searching for someone in the crowd who has three or four minutes to waste with them, someone who surely must be important to oversee all this variegated bustle, someone who is not a worker in dirt and mud but perhaps a manager or supervisor after all, knowledgeable of computers and cellular gadgetry, perhaps someone, given different circumstances, who might have been not so different from themselves.

When Kobashi signed us up to deliver raisins, we had no time for him, even though he was a rare ex-farmer himself and had an inherent *dignitas* characteristic of the penultimate farming generation as it neared seventy. We did not go to lunch with him. We did not follow him to his car. We did not even wave when he drove by on his errands. We felt awkward enough each year when he drove out, walked into the yard, and of all things ordered us, us on our property, onto our own lawn, to fetch a few chairs under the tree, and sit with him that we all might formally talk raisins, the bleak, boring tale of the marketing, the selling, and the disposal of dried grapes.

Gradually, despite this formal and quite queer ritual of hospitality, Kobashi won all over, well before his dramatic appearance before the collapse of his own employer. He was not overly pessimistic, as many are who know the sordid inside of the industry. He did not revel in etching a vast outline of doom, of oversupply and shot markets, of jittery banks and new imports, all to shock the dense, blinkered farmer who trucks his raisins in each fall, who simplistically believes that right then his work has ended, as if those towering stacks will mysteriously shrink when he returns twelve months later. The naysayers and storm clouds pontificate to and patronize the vanishing farmer, as if their secretive tales of an agrarian Armageddon put themselves above the fray, as if the vine and tree man had not himself tasted something of this agricultural calamity.

Nor did Kobashi go to the other extreme, in the manner of the more ignorant and inexperienced field men and salesmen. They are the spinners of tales, those seers of a prosperity just over the horizon, replete with the constant quotes from the high boss, who in turn quotes his more regal and far-off broker maximus. Those aggressive nearly climb on your back, jawboning and flittering from insurance to mortgages to speculation, dumbstruck that the dullard farmer would sit mute on his ancestors' equity when it could provide a ticket for a three-year swirl in real estate or junk bonds. God, they were and are so awful, those gold-chained hucksters whose leather pouches can sprout anything from insurance to annuities to brokerage and real estate deals. No, no, absolutely not, you tell them, as the infectious agents are shepherded away from your unvaccinated wife and children and stuffed back into their cars. Countered and checkmated, with a last gasp they roll down the window to make a pitch to your pragmatism with more mundane eucalyptus farming and crawfish ponds.

Not so Taro Kobashi. As is perhaps the custom with the best of the elderly California Nisei, veterans of the camps, the forced train ride, and the stolen farms, he said little. That little Kobashi did say always made some sense. When someone is wrinkled from years on the tractor and when he is hardened from the brutality of the regimen and the coarser men with whom he must deal hourly, it is altogether odd to find in him a quiet, almost timid voice, manners near urbane, and a vocabulary at once free of both obscenity and pretension. And so at times we did talk raisins with elegant, soft-spoken Taro Kobashi, and we learned about what happened to our harvests before they reached you.

Aram Manookian down the road, on the move, always onward and upward with the new vineyard stakes and big shed? Kobashi cordially laughed him off. "I wouldn't be impressed, boys, he's doing nothing different than you, averaging his two tons of raisins an acre like everyone else. If anything, he loses more than you. I guess you didn't have a mom with real estate down by Sea World to leave you a million or two. Now can't anyone farm pretty with a million-dollar cushion?"

Was anyone else as bad off as us? Hoping to hear again and again what we knew to be true, we pressed, grasped pathetically for a field man's anecdotes of nearby agrarian ruin and bank foreclosure. "Your pump isn't locked up yet by PG&E, is it? What would you do if your vines needed water and the power bill was overdue? A lot of pumps are padlocked right now. A lot won't be unchained for a while. It costs between $700 and $800 a ton to get the raisins out the yard, so when you're getting $500, the bigger the vineyard, the bigger the loss. I'd have to say it would be better for all that those pumps would stay locked, then maybe we'd have less vines."

The infamous, all-powerful Larry Black, with his helicopter and company Cadillac, who was in charge of the vast monolithic Valley Girl Raisin Cooperative in its darkest hour? Would not Taro at least speak in defense of a fellow raisin manager? Not at all. "Black's team grabbed a hold of as many raisins as they could, and then they had to sell them for any price they could get. The plant lost money every hour it was open. Damn the price, damn the farmer. He supposedly took his bonus on tons moved, and he made a lot of money, over $300,000 a year, I hear. It seems to me it was always all for Larry Black. Isn't that wrong? Are you running a cooperative for your members or a private company for yourself? Who needs the car, airplane, and house when it's coming right off the trays of the farmer? So what if he gave his management team big bonuses? Nothing's worse than being generous with someone's else's money."

Kobashi uncharacteristically showed a little temper and loquaciousness when the subject stalled on the former Valley Girl Raisin Cooperative president, as if Black had darkened the reputation of all (like Kobashi) who handled raisins once they were forked off the farmers' trucks. Kobashi couldn't stop, in fact. "Why, Larry Black laughed at the idea of RID. [RID was the Raisin Diversion Program, a farmer-financed self-help plan where any who promised not to produce a crop were given partial payment in raisins by their fellow growers.] He wanted everyone to produce more, truck them into him by the semi-load, so Valley Girl could sell even more of them at a loss."

His own boss, the backwater corporate speculator who had taken over Central Valley Raisins, Incorporated? Embarrassed, Kobashi didn't say much at all good. He pointed out only that the plant was close by, the trucking would be cheap, and we would get paid about the same as anywhere else. Indeed, eyes rolled when he finished off his sales pitch—more a sermon than anything else—by matter-of-factly announcing that he of all people, five-foot-four-inch, 140-pound, seventy-year-old Taro Kobashi, would guarantee that the checks were good. There was a near arrogance in his self-assurance that he would not warp and that we, of course, were to believe he would not warp.

I envisioned the outcome rather differently: When our first twenty tons of raisins would be red-tagged for nonexistent moisture, rocks, sand burrs, eucalyptus leaves, or rodent feces, would we not perhaps see the needed Mr. Kobashi hiding by the weight scales or unreachable in his company car as his employers docked the load at $30 per ton?

Taro Kobashi, despite his dust-free automobile, his impeccable shoes, his immaculate but tasteless polyester, was the first and last field man that we ever could like, for he soon would have his small moment to keep that very pledge and more. One August he gave us a ring.

"Heard anything about Central Valley?" he asked.

"Yeah, a little scuttlebutt."

"Don't listen to any of the rumors you hear about us in the next week, good or bad. I'll see you in two days when things are clearer," then he hung up as abruptly and mysteriously as he had called. My brother called the pack together, and like rabid dogs we swore to each other, vowing that little Mr. Kobashi who signed us up with these crooks would be boned and shredded on arrival and sent back to Central Valley ground and canned.

Precisely forty-eight hours later as promised he drove in and walked erectly up on the lawn. In measured tones he quickly outlined what he only recently had surmised about his employer through rumor, hearsay, plant gossip, and flustered denials—absolutely accurately as events turned out. We were to be quiet. We must follow his instructions to the letter. He would be fired, no doubt, laid off, or possibly even sued for telling the truth. Nevertheless, we were to do exactly as he said. "I don't want anyone to lose his raisins when it doesn't rain. Wouldn't that be something—to hand your raisins over and still lose your crop?" He laughed nervously. There was not much worry or doubt in his voice nor a wasted word, much less anger or recrimination toward his employer.

The company, his company, Central Valley Raisins, Incorporated, was, as most thought, about broke, he reasoned. Most of its profits, all of the efficient work at his plant, were used to pay junk-bond debts out of state. (This raisin packing was a very lucrative business. After all, those in between the grower and the eater were acquiring raisins now at barely a third of their 1982 field price to farmers while the price to the consumer remained about the same.) The owner of Central Valley, Kobashi guessed, lacked the capital and probably the credit line to pay for delivery of anyone's raisins. Nevertheless, we were all under contract to deliver. When that happened, either the banks or the company would seize the raisins as chips in upcoming bankruptcy proceedings, the farmers' payment relegated to the bottom of the creditor list as "unsecured debt," meaning weeks without the wherewithal to pay the bank or, worse, possible forfeiture of any recompense altogether.

We nodded, adding that stern letters had in fact been sent out to growers just today from Central Valley, warning them to follow their contracts, ignore rumors, and bring in their raisins. Kobashi had apparently already had his run in with the company about the need for honest, full disclosure, been reprimanded, and now was to choose between his job (and reputation as a loyal company man for any future employment) and the farmers whom he had signed up. Since Taro Kobashi had no hard proof of either dereliction or malfeasance, I still suspected that he might simply be announcing to us all his suspicions, alleviating a little guilt in the process, only to promise that things would probably turn out all right in the end once we hauled in our raisins.

Not at all did Taro Kobashi do this. No Hugh Effin or Buddy Messingale was he. Instead, he lowered his voice, hunched over, wary that our wives or kids should not be distressed over his melodrama. "You can get out of this little jam if you do what I say. Read your contract carefully. You are entitled to ask for a cash advance for your harvest costs. If the company doesn't give you some money, you can break the contract. Clarence Hardwell and Yip Nakumo did it and it worked. They're out of Central Valley just this morning."

Were we really to believe that Taro Kobashi had been busy undoing his own labor of the last few years, educating all those stubborn farmers, whom he once had so diligently signed up, about their legal rights outlined in their contracts? Who was this strange anti–field man, giving up hard dollar commissions in exchange for the tender of good will and fair dealing?

Within hours, despite the threat of lawsuits, despite the letters of denial, we called Central Valley and for the first time in our lives asked for our cash advance. The stunned appendage at the desk begged off and tried to redirect us to his boss. Finally, from the head hireling himself we got the expected negative answer to our capital request with a variety of the excuses common now in the California fruit business: "The money is tied up; we didn't budget you for harvest costs; we'll get it to you in a month or so; come in with financial forms," and so on. Immediately we drove into town with the Kobashi-supplied releases, ourselves "enraged" and now threatening to sue instead of being sued, lamenting the loss of "critical harvest funds," reminding them of the vicious rumors of plant foreclosure in the papers. "We demand," we yelled, "to see Taro Kobashi, our field man, to get to the bottom of this so we can get the cash to pay our pickers."

By five that evening we were out of Central Valley, and Taro Kobashi was not much later out of a job, parting with the company that was betraying its own growers. Despite his ostensibly Asian obeisance to authority and his antiquated sense of loyalty to his boss, Kobashi somehow had, as so many of the Nisei have, fashioned a code of conduct far superior to anything Eastern or Western now left in the valley. He had learned about the value of Western independence and resistance to bureaucracies from the native valley yeomen now deceased (and no doubt from his own deportation and internment); the politeness, deference, and sophisticated savvy were all Nisei. I suspected without the slightest shred of evidence that he was a far more complex man than even his noble, long-suffering first-generation-immigrant father and a far better man, too, than his own Westernized offspring.

It is regrettable that America cannot freeze, clone, and store the second-generation immigrant, who inherits all the drive of the first, who adds kindness and honesty to his wearied father's harder edges, but who misses out on the leisure and affluence that ruins the third. From Aristides to Pericles and on to Alcibiades, it takes but seventy years and it's all over. Is it not also regrettable that the laboratory of individualism and skepticism, the American agrarian, is also finished? From Italian yeoman to Roman banqueter, from the doomed at Cannae to the urban mob waiting at the dock for the weekly dole, was but two centuries.

Kobashi was no zealot; no Antigone was he looking for a doomed cause and a write-up in the Fresno Bee. On the contrary, Taro Kobashi was genuinely ashamed of his small company, offering us the plea that they were all good men who must have gone over the line in the bond craze of the 1980s. Good men they once were, who now would take their profitable, hard-working plant and suck it dry for losses incurred on a computer. He was genuinely sorry for wretched them but apparently more sorry for what might become of wretched vineyard men like us.

The first owner drank Central Valley up, Kobashi sighed philosophically; the second gambled it away. There would be a third, some day after he was gone from the business, with more capital and better managers, for the raisin-packing business is a very lucrative enterprise.

So there are raisins once more being packed at the old Central Valley plant. The antiquated machinery is now refurbished; the yard reblacktopped; new, smarter suits lured in, as a national food corporation has purchased and reorganized Mr. Kobashi's old employer, Central Valley Raisins, Incorporated.

Oddly, Kobashi never said much after our crisis was averted, neither lamenting the loss of his job nor preening for a few crumbs from us about his heroics. Heroics they were, for had we delivered that August and lost that crop, even had the payment been delayed, we would have gone under right then and there, and there would be nothing left to rent out, fight over, mortgage up, or sell off as I write today. When Taro Kobashi, now retired, drives in during the late fall, anyone who is here walks over, draws up chairs, and sits under the tree with him to talk raisins. They've never said so, but they like intensely this strange man, this Taro Kobashi, this tiny savior of men's lives, quiet protector of the ignorant agrarian, this odd polyester man who cannot lie.

II. Ulysses Ponce

Ulysses Ponce was a most bizarre labor contractor to find in family farming, to find anywhere in agriculture. His speech was part anxious stutter, part aggressive shouting, flavored with both a southwestern and Mexican accent. He was apparently a barrel-chested, swarthy Mexican-American (if he was not Mexican-American, what else might he be?), who presented with symptoms of no discernible age, race, or occupation. He was somewhere between thirty-five and sixty.

Do you shudder when you hear the title "labor contractor"? Men like Mr. Ponce usually draw only disdain from farmer and farm laborer alike. They earn outright hatred from the urban reform movement, which targets the convenient man in the middle between producer and picker, the contractor who cannot tap into the easy and cost-free American nostalgia for either small farming or its honest wage labor. Yet in the nineteen years we have known him, in the nineteen years in which he has ruined his back, gone bankrupt three times, lost his job, lost his ranch, lost everything but his house, found and lost his god, he has never lost his respect for the men he must employ and his interest in preventing the fruit of the small grower from falling to the ground. He possesses a competence and bravery far superior to men much richer and better bred, this perpetually optimistic creature who failed so miserably all throughout the 1980s. Lest I forget, Mr. Ulysses Ponce has never taken money that is not his. And as I write, he is near destitute once more.

Ostensibly there is nothing good about the labor contractor. He is the punching bag of the union labor reformer, the Mexican nationalist, the conservative farmer, and the state labor and occupational regulator alike. Does not he corral the homeless, the weak and the poor, to bus them to the fields of the absentee baron where, without adequate sanitation and water, they work below the minimum wage, their pay scarcely paying for the contractor's forced deductions for exorbitant transportation and food? Sometimes, perhaps very rarely if at all now, this can occur on the west side, the corporate side of the San Joaquin Valley.

Is he not the subject of the TV documentary and newspaper exposé, traitor to his race and class, the capo who lashes his own before the greedy landed and exploiting class? Does the contractor not pay his men fifty cents on the dollar so that the other half might go for his retirement rancho in Mexico? This, too, is not always the stuff of romance and fiction; Raul Chamorro insisted that all his men sign up for welfare so that he could pay them only half of their earned wages (in unreported cash). Between their halves from the government and Mr. Chamorro, his "unemployed" crews might still get a full paycheck. These like Raul Chamorro were the cowboy-hatted, stylish contractors with Viva Reagan stickers on their truck bumpers, which incensed so many. To the radical brown they were betrayers of their race, only to be labeled as upstart wannabes to the white reactionary, counterfeit emulators of the better-bred pale looter who scored in RV sales or probate law.

Is the contractor of labor finally not the petty embezzler, the government cheat who steals the social security and workman's compensation from his men's checks to pay for the diamond on his finger and the $30,000 luxury Wagoneer with the cellular phone and leather seats? Yes, men like ChiChi Martinez did all that and more, their empires once reaching two thousand men in the field, involving trucks, transportation, land, barracks, bars, restaurants, emigration and accounting services, gold, jewelry, cars, and homes, before the arrival of the federal agent who shuts it all down and schedules $500,000 of fines for the rest of his life, deducted from his new job behind the wheel of a delivery truck. It is especially hard now for ChiChi Martinez to labor for one boss when so many hundreds have labored for him; particularly humiliating to be descending back to his origins when he had the game won in his thirties.

Ulysses Ponce was none of the above, an honest man in a disreputable business in a dishonest age. His only trappings of success were an immaculate blue baseball hat and a shiny rehabilitated El Camino pickup. He wore thin white starched shirts over his sleeveless T-shirts, the smell of Brylcreem all doused over with a heavy dose of very cheap aftershave. Those were his medals because that was all that he might squeeze from either laborer or farmer, who in turn much more often both gnawed at the bridge between them, the solitary upright Ulysses Ponce. Because he might not lie, could not steal, and certainly would not loot and maim, he made little during the boom of inflation and lost that little during the postcrash years of the eighties when men, many of them friends and family, in desperation stole and looted from Ulysses Ponce himself.

Wily contractors of daring and brains, like Ulysses Ponce, ChiChi Martinez, and Pepe Ortega, bring crews of ten or fifteen men in each day to pick peaches, plums, nectarines, and grapes. They supply battered fiberglass toilets, water cans, and transportation in enormous cargo vans, with two long wooden benches nailed to the floor. They pay the men $5.50 to $8.00 an hour, depending on whether the farmer and crew agree to a piece or hourly rate. Contractors pay $7 an hour to a crew boss for about every twelve men. Usually the best of these leaders is a veteran of thirty or so campaigns, with shot back and bum knees from going up and down the ladder with forty pounds of tree fruit strapped to his belly. Lago Gutierrez and Rigo Santayana were Ponce's signors, fifty-five-year-old Lago carrying an enormous three-hundred-pound stomach that had won himself a new twenty-five-year-old wife. She was a young Chicana still stunning after a decade of heroin, still in perpetual motion that was soon to rejuvenate, then tire, then exterminate Lago Gutierrez.

The bosses drive the full bins into the loading yard, refill the water cans, supervise lunch and two fifteen-minute breaks, and monitor the color and quality of the fruit as it is picked. In Ulysses Ponce's case, the field bosses were neither the surly preened to hammer the men down nor the gang emeriti to frighten the farmer with block-lettered insignia tattooed in their skin. They were like Ulysses Ponce himself, played out, too talkative, too kind to worker, too sociable with farmer, too nostalgic for the 1970s and the old paternalistic system spawned by a once-generous inflation.

Wages are paid by the contractor, social security is deducted and matched, workmen's compensation is subtracted and matched, and all immigration forms are xeroxed and kept on file. At work's end the contractor presents the farmer with a bill to cover the wages, social security, workmen's compensation, and his own profits for assembling the crews; that can run from 28 percent to 36 percent above the cost of the actual wages paid to the men. Usually the tab is not computerized, but rather simply written out in longhand by Ponce, with wordy explanations for various charges when the sum is especially high. Apologies are made by Ulysses Ponce if he thinks he is costing you too much money, when his used El Camino looks better than your aging Mazda minitruck. Usually, however, you must spot the error in the arithmetic and politely tell Ulysses Ponce that he is due more money than he is asking.

Within that system graft, illegality, and cruelty dominate, for the man in between must both satisfy and profit from those on his poles, and the money that changes hands from bank to farmer to contractor to worker is in the many thousands each week. Some contractors, for example, charge two dollars a head for the ride to work, and they can also make a lucrative business out of delivering to their own stationary men ice-cold Pepsis in the middle of a 120-degree peach orchard for two dollars a can. The less bright test the never-to-be-tested IRS and for a time, usually no more than two or three years, manage to pocket their crew's social security and disability deductions. A few simply round up, for the price of cheap alcohol, derelicts and criminals who cannot work and will leave fruit on the tree or on the ground, anywhere but in the bin. The smallest minority of labor entrepreneurs, the most unabashed and clever, pad their hours and bill the farmer for work that did not take place at all. Their bill reads: "Monday July 8, 22 men picked Santa Rosas, 9 hours, crew boss $7.50 an hour," when, in fact, the men finished early and went home at noon. When pressed and cross-examined, they simply retreat to their pickups and shamelessly write out on a notepad an adjusted official-looking revised invoice.

For their part, workers may or may not show up for work, leaving a crew short when the fruit is overripe or bloated and squabbling for work when hail has reduced the picking season 90 percent. Other laborers, worried about random gang violence at home, bring their children to run wild but safe in the orchard and let the contractor and the farmer worry about labor violations and the government's concern over the presence of "oppressed" underage children wandering among shady and secure trees and vines. Many pruners, thinners, and pickers turn up sick or infected and need immediate medical attention. Apparently a rotten tooth or a spreading gonorrhea infection may get a worker off a ladder ("work related") and into the clinic quicker and safer than if he lies up in bed at home. Many more need cash immediately advanced against their wage and upon receipt fly south. Nearly all have false immigration papers and social security numbers and so demand hiring or else; the felonies go to Ulysses Ponce if the documentation is found forged. A few are the no-good, the real criminals, the rapists, murderers, and thieves on the lam from southern Mexico, who can bushwhack the contractor and take his always wide wallet of the cash needed for his gas, drinks, and advances.

The farmer himself remains the ultimate arbiter of the system's success or failure. If he can simply give a contractor like Ulysses Ponce two or three days' warning of his needs, the contractor can accommodate. Too often, however, the farmer himself does not know whether it will rain and cool off or be clear and sizzling the next two days. Many agrarians who defer to the broker, who meekly obey his bark to wait three critical days before picking, who nod their head and stare at the ground when the company representative suddenly demands twenty pallets the next afternoon, themselves transform into foaming tyrannosauri in hours.

The now-desperate growers ring their contractor up at eleven in the evening, yelling, swearing that they need men, many men, and they need them now! They threaten to end that night ten years of business with a Ponce or Ortega; they vow even to go to ChiChi Martinez or elsewhere where the men can be found; they resort to anything to find the men, to get the fruit to the broker when he says he wants it. They can humiliate their contractor, their only lifeline in their panic. They do and will do anything to get the men and save their year; always the men they search for when the fruit is ripe and a truck heading east is idling empty at the dock.

Just as often in a slow year, the plum and peach man with no memory avoids the contractor and his hordes of hungry men who have no work. When a frost, a hail, a light set has ruined the grower's year, the farmer reasons correctly that he has no money, so he must cut himself off from those who have even less. When there are many arms and scarce fruit, he drives a hard bargain over wages and hours, the beleaguered farmer does. Just as he has the year prior expected Ulysses Ponce to have six, seven, and ten crews at his disposal to meet the huge harvest, so now he is disgusted that Ponce has set aside two or three crews for him. "Sorry, Ponce," he says, "There's no way I need all those men. Try me next year." As I said, there are no longer any Rhys Burtons who toss and cannot sleep and so the next morning call back, "Bring them over and I'll find something for them to do, you were pretty good to me these last two or three years." How could there be when orchards are no longer of two hundred and five hundred trees but of ten thousand and more, when the computer and the chain of command alone dictate what is and is not productive labor?

The worst fear for Ponce is the farmer who is nearing bankruptcy himself or who is dishonest and never pays until he himself is paid. They reason, why pay the suitless, tieless Ponce of all people when you are going under? Why borrow money to pay sorry Ponce, the bad-backed stuttering nobody in the old El Camino with his gapped smile, when he can wait for your check and pay the accruing interest himself? This cheat is rarely the grower of twenty to a hundred acres but more often the more flamboyant would-be big man of three hundred to one thousand acres. If he has five hundred acres of tree fruit, sixty thousand bearing trees, he may need eighty men each day for three months, creating an enormous problem of logistics and accounting for the contractor but also a source of real profits or instant liquidation in one fell swoop.

Ulysses Ponce is not the type of man who can walk into the Bank of America and demand a million-dollar line of credit to pay his men until he too is paid. The farmer who stiffs a Ulysses Ponce one time puts him out of business right then and there: this happened three times to him between 1980 and 1991. Even if Ulysses can buy a reprieve from his workers until he begs the failing grower to cut a check from his loan officers, he is still not safe from two hundred working men who think it is not such a crime to slice Mr. Ponce's jugular, to slay at night Mr. Ponce for putting them out in day amid the peach fuzz, the heat, and the dirt when he would not or could not pay them their due. Like the bank, the workers care little to hear Mr. Ponce's *principia philosophiae.*

In the years since I first met Ulysses Ponce in 1976, he has thrown out his back picking fruit himself when his crew failed to show up, paying huge sums for chiropractic, elective back procedures, and routine uninsured office visits. He lost everything but his house when Tal Carter, the pompous nectarine and peach czar, used him for a month and then failed to pay him $50,000 in services rendered, when the market crashed from twelve to four dollars a box. Two years earlier, Ralphie Bettencourt stiffed him for $25,000, as Bettencourt was liquidating his assets, turning his debits over to an insurance company, which saw no purpose in paying Mr. Ponce for fruit that was no longer there, for men who had long since disappeared, for work that had no value or repute.

When Ulysses Ponce crawled out of his abyss, reorganized, and once again soldered the wire between the farmers and the pickers, Jorge Salinas was waiting for him. This not-to-be trusted bookkeeper, the all-in-one local accountant, immigration fixer, money lender, and bail bondsman stole all Mr. Ponce's social security and disability deductions. Mr. Jorge Salinas left town with the checks and has not been seen since. I met Jorge Salinas once during the 1982 harvest rain when he came over with Ulysses Ponce; he offered to do our books free for the first six months.

The only profit left in the farm-labor business in the early eighties was through some sort of assorted theft. When prices fell from eighteen to eleven cents a raisin tray, when hourly wages crashed from seven dollars an hour to $4.50, commissions in turn were cut, checks bounced, and trees were left unpicked, unthinned, and unpruned. The government took Ulysses Ponce's small orchard, his own entry into the world of his employers, as he paid them by the month for the not-to-be-found Jorge Salinas.

Ulysses Ponce staggered back from the dead for a third go. "Look at all the trees around here. They have to be picked. People eat. I can get the men to do the job. Not many else can do that the right way. There's always a place for me, if I don't get stupid and hire a Jorge Salinas or lend money to my brother-in-law's restaurant."

You, any of you, can still today call Ulysses Ponce, and in desperation say: "Ulysses, I have a problem. I need thirty, not twenty, men tomorrow. They have to color-pick, not strip, the Elegant Lady peaches. And I need at least thirty bins from them by four o'clock." Ulysses Ponce answers, "You'll have twenty-five men with Rigo Santayana at six o'clock. He'll meet you in the orchard, and they'll be good, fast pickers and you'll have yourself thirty-five bins by three-thirty."

But it will be Ulysses Ponce himself who will lumber out of his El Camino in the early morning dark, wade stiffly into your field to climb the ladder and show the men exactly what he, Mr. Ponce, wants done. He will also see you on the alleyway and storm over as dawn rises and the weather warms. Ulysses Ponce will commiserate over the poor price of peaches you are receiving, and he will brag on the dexterity of his men. When you get defensive that he might be insinuating you cannot pay his professionals with those meager prices, he apologizes profusely. He says he knows that you would rather go broke than stiff him and that's why he brings you the men this morning, every morning. He is right. Most farmers who know Ulysses Ponce will do anything rather than not pay him and his hardworking crews their money.

In May Ulysses Ponce will be a charter member of the Free Will Evangelical Ministry. He will look you in the eye, stab his crooked finger into your chest in your orchard, and tell you with squinted eyes that all the farm ruin is God's will to test men like us before we die. He reasons low prices and shattered lives present a wonderful—the only—chance to win our souls' salvation. The next June the now-pagan Ulysses Ponce will proclaim that he can cure his back and the damage to his back from chiropractors, *brujas*, and seedy doctors only "with the shovel digging weeds." His mushy prostate, abscessed teeth, hemorrhoids, high blood pressure, high blood sugar? They are minor annoyances for a man of action. Government or no government, he feels like he's twenty again, and he's now ready for a third time to buy land and become a farmer of plum trees. The July after, a Catholic and near cripple once more, Mr. Ponce will roar in the driveway for consultation, advice, and solace over bankruptcy papers from a failed plum grower. He will say in desperation that in the end his delinquent, no-good, broke cheat will, after all, pay him, Ulysses Ponce, and that the bank's papers are just a formality (he did not pay Ulysses Ponce). By the final crop of the decade, when the religion, the back, and the money are all gone again, Ulysses Ponce preaches to you that he doesn't like Reagan. Reagan, alas, is for the wealthy, he says. But Reagan is also good for us, he sighs, because our president alone understands that we, all of us, you and me, are not at all good.

You never patronize, much less dismiss, deprecate, or ignore, men like Ulysses Ponce, the anonymous engines that drive the entire California fruit industry, whose cunning and endurance bring America its sterling produce. Even his underlings, Rigo Santayana and the late Lago Gutierrez, understood man in his physical universe far more clearly than your glib colleagues in the school of social sciences who now teach jammed classes in self-esteem and the theory of walking, both nontopics that these two crew bosses thirty years ago mastered. You listen not to the university denizen but instead to the lectures of the near sixty with missing teeth and scarred hands, who get up in the first light and send some pretty formidable men over hundreds of miles of vineyard and orchards, who do business with a handshake and do not take someone else's money when there is so much money of others to be taken. So I learned far more from the bitter proseminars of Lago Gutierrez and the colloquia of Professor Ponce than what little escaped the adolescent brain of Dr. Leslie Doily, the pundit of the university humanities program. If there were but a shred of justice left in this world, if the campus were a place to discover truth and to acquire courage, veracious, gap-toothed Ulysses Ponce would now be the prized chancellor of the university. The president (himself, like Ponce, without the terminal degree) in turn might better belong as the twelfth man on Rigo Santayana's twelve-man peach-picking crew, a bunch not much interested in his TQM, distance learning, and diversity administrative apprenticeships.

Such imagination, if unrealistic and immature, still is not quite perversity. If Ulysses Ponce looks you in the eye and sees some good, if Lago Gutierrez shakes your hand for a fair deal all around, and if you can drive in vineyard stakes all day beside Bus Barzagus, you feel how hollow is the E-mail thanks from your fellow professor on the general education committee. You discover why you never found delight when the provost paused for a few words of pro forma chitchat at the Christmas reception. You are doomed to a life of cynicism when a tweedy grandee from a private university nods your way, when the self-important journal writes that they have tentatively accepted (with four pages of prerequisite changes) your bizarre article on the runaway of ancient Athenian slaves.

You wonder also about these better men of a better time, who still lurk about trees and vines where for fifty years they have lost their youth and health picking and pruning for little pay. I ponder now how and why—when I see that the Princeton-educated will loot and manipulate, when the San Francisco lawyer pads his bill, when the Stanford humanities professor will schedule two classes at once, when the Berkeley scholar on sabbatical complains about an exhausting (but wasted) three hours at the computer—that broken-backed, perennially failed Ulysses Ponce is never bitter that he will not take money he needs when it is all around him for the taking, is not concerned that he is a better and poorer man than the professional class that runs his country.

"There is no money," Ulysses Ponce declares when pressed about an undeniable connection between his failures and his honesty, "it's only people. You grab the money you need; then you're stealing from a guy or his wife and the kids. I steal from you or the men, and then how can I get mad when someone stiffs me? Me? It's not for the money anyway. I like the work, I like to see men working, picking fruit all day, everybody outdoors in the trees. Without us, would people eat? I like the shovel in my hands; it clears my head. Next year I am going to buy a new pickup, with two seats, back to back; Lago, Rigo, and I are too fat now to fit into the El Camino."

I saw Ulysses Ponce recently; he was now driving a white double-seated Chevy pickup, used but in apparent excellent condition. He was complaining that the frosts were late, the rain scarce, the leaves still on the plum orchards, and so his tree men were idle but eager to begin fall pruning.

III. Rollin Buckler

Still, there are many others in the pantheon beside Taro Kobashi and Ulysses Ponce, many farmers who walk their ground in silence, who timidly disk their vineyard until fate drops a spark on them. When that happens, beware; for every good thing that the art of viticulture, whether it's labor or management, can teach a farmer—patience, wisdom, piety, and toughness—they possess tenfold and more. These loners are not afraid of capital and the power of capital. They care little for the word of other lesser but wealthier men in their midst. You see, cut off from others, king on their tiny domain, alone on their tractors, a few select think and brood, not over their standing in town, nor even about the size of their acreage and bank accounts, much less about cars or women that might yet accrue. They need no periodical retreat to sort out their lives, for their lives themselves are one permanent retreat. In times of emotional crisis, when it is understandable to question one's existence at forty, they do not panic, divorce, inhale drugs, much less color their hair or buy a scarf.

More often they might dig vines a little slower, a little longer run their water, lean on their shovels a little too frequently. They stare as the grape clusters bloom, mature, and sugar, the leaves sprout and fall, and they develop themselves a harmony, a rhythm they find themselves in with the water, the heat, and the cold, a rhythm that suggests all others should be like themselves, all with their muscle, brain, and nerve should likewise clock themselves into their own lowly station, their small link in the bigger chain of growth, death, and renewal. They do not necessarily see their tiny domain as small. For even sixty acres of vineyard translate into thirty-six thousand vines, and that is quite a kingdom to prune, tie, water, and pick. Why, out in the middle of thirty-six thousand vines, the farmer can find enough physical universe for a lifetime and more. Even from working something as mundane as Thompson seedless grapes, some cultivators become utopian visionaries and imagine (quite theoretically only) the way things could and should be on the farms and town around their own orb. They read no book; they hear no sermon; they sit in no class. And yet they acquire wisdom about man from nature, there for any to tap should they choose isolation over security and comfort.

So far, so good; for the undiscovered heroic have no belly to ride out to protect the poor and the meek, much less to become one with the now-disreputable urban crowd whose stock, trade, and con is the protection of an anonymous and purposefully distant underprivileged. These unknown and undiscovered farmers occasionally pull on but do not break their leash.

But if it is cut? If it is severed by someone else, especially someone who seems foul and odious, who does harm, manipulates, and bankrupts the agrarian calling? They will lunge off their tractor. All incisors, the pent-up energy of years of forced contemplation fuels the attack. These nondescript dirt shovelers turn furious that others in fact are not content with their meager share of the cosmic process and have taken from the agrarian his own very small due. Men in plush Japanese and German cars who phone as they whizz by are no longer the mere misguided. No, they and their riches are now proof in the flesh of their greed and they must be dispatched to the lower rungs of the inferno.

This revelation of raisin mismanagement mobilized of all people forty-five-year-old Rollin Buckler, third-generation agrarian, vineyard man, former law student, past linoleum salesman, accountant emeritus, once husband, divorced father, and inconspicuous resident of the trailer-house pitched behind his mother's and father's small farmhouse. Occasionally overweight, bearded and slightly balding, Rollin Buckler became the arch antagonist most unlikely of the multimillion dollar Valley Girl Raisin Cooperative.

When Rollin Buckler left his trailer house, borrowed his seventy-five-year-old father's pickup, and went to war, everything he had been unable to make much money in—accounting, business, farming—instantaneously was transformed. They were now the rare, absolutely essential prerequisites, the only arsenal of experience diverse enough to match a cadre of bought professionals. It turned out that Rollin Buckler had *never* failed at all!

No, like U. S. Grant back in Galena, Illinois, Mr. Buckler had simply been unknowingly engaged in years of careful, unrequited graduate study, all to ready himself for his epic moment against the hired minions of the Valley Girl Raisin Cooperative. The problem with Rollin Buckler, in Aeschylus's words, was that "he did not wish to seem just, but to be so." And that in America has now become a rare and dangerous thing.

As Valley Girl came to learn, Mr. Rollin Buckler did indeed experience a midlife crisis at forty-five, but self-contemplation presented not in permed locks and pierced ears. No, it surfaced in a manic desire to make the antiagrarian world for once follow the rules of his own tiny vineyard.

I became acquainted with this courageous and astute agrarian right after the great raisin crash of 1983 and the subsequent fall of the House of Black—the dark days of November and December, when the membership of the Valley Girl Raisin Cooperative lost their retained capital owned in the association, an aggregate figure estimated at $27,000,000. Despite the resignation of Larry Black, despite the loss of millions of raisin growers' capital, despite lurid tales of waste and rumors of worse yet, the one thousand seven hundred remaining small raisin growers had closed ranks, taken the bullet in the brain, and were prepared to march on to doom with the new, the resurrected cooperative under the leadership of Black's surviving lackeys, who had eased their vile master out.

As Rollin Buckler explained to me, there is a certain stupid pride of the agrarian class in sticking by a bureaucracy, a company, a political organization that has left them impoverished—as if honor, even a certain machismo, is found by pronouncing, "They have ruined me, but I am loyal, no dissident, no rebel am I who would whine and complain when pinned to the wall." I agreed, for I had seen such men daily. But I added to Rollin Buckler that soon most of these robbed would, like Aristophanes' Acharnians, dislike Rollin Buckler far more than they did Larry Black. They did. Nearly all of them.

Two-thirds of the broke raisin families stayed in Valley Girl cooperative after they had lost their money. Remember, the cooperative is never the harbor for the calculating, hard-nosed raisin men, the keen survivors who will swim defiantly alone and endure the rips and chews from the predatory sharks of the private raisin school. There are no Mohinder Kapoors or Vaughn Kaldarians in the cooperative cloister. They are not among the gullible, timid, and idealistic farmers willing to pool their resources, to sell in bulk, and to divide the sales returns so that all might profit according to their station. "Valley Girl," Kaldarian told me, "has too many stupid people who don't need the money, who inherited, never bought ranches; so they'll let any fast talker steal the whole show as he pleases. Who wants to join up with a dumb bunch like that? You're only as strong as your weakest link, and there's some pretty weak ones over there."

In turn, Valley Girl didn't much welcome a Vaughn Kaldarian, an Armenian loner who lacked the prerequisite smug pedigree of its preferred fourth-generation, eighty-acre members, presidents of Lion's Clubs, organizers of the raisin and band festivals, chairmen of the school board, hospital board, irrigation board, and land bank board, the whole dark *cursus honorum* of small-town rural America.

Gil MacMichel, Danny Dungren, Horton Pimental, and Val O'Brien had all inherited considerable acreages and made money in the 1960s and especially the 1970s, wisely if unimaginatively buying long-term T-bills at fifteen percent and more as their father's three-hundred-acre vineyards gave them a steady two tons an acre at $1,300 a ton. Under the insidious Jimmy Carter one ton was cost, the other pure profit. Before they helped to throw the despised president out, these grandees had put away in their banks about a million dollars of raisin profits in the space of his four years.

When these directors of the cooperative, these soon-to-be archantagonists of Rollin Buckler, filed into a Valley Girl meeting and made their way to the front table, all of the brethren were to feel a certain awe and envy in their radiant presence, Plato's timocratic men in the flesh, rapacious for more honors and more land. For their part, they were good method actors all: stone-faced, dour, adding an anecdote here and there about their own legendary parsimony back on the farm, quibbling over a vice-president's free lunch or two in the monthly receipts of the cooperative as they permitted millions of unearned dollars to end up in the accounting books, as Larry Black courted, praised, and stroked them all. (His talk and cant among the board went something like this: "Anyone seen Horton's vineyard lately? It's three tons for sure. I'd better watch old Val over there; he's a tough hombre and I can't seem to get a penny by his nose." Or perhaps, "Gil's right: we're going to spend what it takes, to make this cockeyed outfit first class.")

Rollin Buckler dubbed these tightwad overseers on the board the biggest spenders of other people's money, the most radical gamblers, the most liberal check writers and money borrowers in the history of Valley Girl. For men who felt tough in paying only the minimum wage to their foremen, that was a bitter pill indeed. I saw Mr. Heine once, himself an ex-board member and apostle of Larry Black, and remarked that a third of a million dollars for the agora lounger, the *polypragmon* Larry Black was quite a tab; a lot of lesser minds, myself included, could have done Black's job for a third of the cost.

Valley Girl Raisin Cooperative was largely the domain of the small-scale success with a paid-up, inherited forty- to eighty-acre raisin farm, who grew up with cooperative blood in his veins. These men remembered stories of the black years of the Great Depression when only their beloved Valley Girl would protect the grower and give him something for his raisins, when nearly all their fathers and grandfathers had joined Valley Girl and so had saved the raisin industry. For the faithful who in the 1930s had volunteered their labor and time at the plant, Valley Girl was a religion, whose creed thundered: "Stay with us and you never shall want; you shall never deal with the wily private packers, the disreputable winery, the crooked fresh shipper. For your mindless obeisance Valley Girl shall protect you who are timid, weak, and ignorant of sales and marketing from all those that might hurt you. We shall not always enrich you, but we shall protect you, for we are you."

Rollin Buckler and I had also been infected with that virus at birth. Rhys Burton, after all, was a charter member of the cooperative. Even my more worldly mother grimaced when she saw Mohinder Kapoor's raisins go by, headed to the private packer. "If we could just stick together," she'd sigh, "we could stop the nonsense and get the price we deserve, if everyone would just be a part of Valley Girl." So said they all, the matriarch Louisa Anna Burton, her son Willem Burton, Rhys Burton my grandfather, us too—a century and more's naïve belief that the most independent, cantankerous, oddball misfits in the world could agree on a simple, uniform principle not to produce unless the price was good, to sell their own produce to the consumer themselves, not hand it over to the middleman, to show the country that farmers were united for more than price, for the greater agrarian principle of staying on the land. In the end, all Valley Girl proved was that when farmers themselves processed and sold their own fruit, the resulting bureaucracy was no better than a Ross Lee Ford or Hugh Effin.

At the height of the lawsuits and countercharges, Rollin Buckler never denounced, as many did (myself perhaps included), the spirit, the noble and original aims of Valley Girl. This Aristides the Just was not to be dismissed as an anarchist, as were the more imaginative, vindictive, and reckless who wanted Valley Girl liquidated, its plant sold off to the Japanese to cover the losses of its membership. He was a reformer, who wanted it out of mergers and acquisitions, who demanded smaller salaries, better accounting, and more efficient workers in this historic communitarian enterprise. Rollin Buckler was not easily portrayed as an arsonist of all his friends, family, and neighbors had built.

So the Valley Girl board and legal team came to learn that he was much worse than a radical environmentalist, union organizer, or federal trustbuster; Mr. Buckler actually was to be a relentless chaperone for Valley Girl, a restorer of her fallen honor, to pledge with his computer, his borrowed law books, and his unimpressive briefcase that she not be compromised and not at all seduced at will. Of all people, Rollin Buckler alone believed in the founding principles of Valley Girl, the farmers' first and last hope, a truly communitarian packing and sales cooperative!

Rhys Burton, as I said, was a founding member of the cooperative. He preached its gospel to us all. He welcomed the idea of agrarian egalitarianism, the depression-spawned idea that hundreds of small growers would pool their raisins, do their own packing and selling, and so split the profits as if they were all themselves the middlemen. I, too, when I began farming, like my mother and father, was "shocked and sickened" by our neighbors who had deserted Valley Girl for the cash up front from the independent packers or multinational corporations. Did they know that without Valley Girl and its heroic struggle there would be no raisin industry? Without Valley Girl would not East Coast corporations and dishonorable locals consume raw the raisin grower, paying him fifty cents on the dollar for his year's work, ceaselessly pitting one grower against another in their greed to drive down the price of dried fruits that could be hoarded for years? Did they not realize that as long as farmers packed and sold their own produce, their return could not be undercut by private brokers, that all farmers could always find shelter and salvation with the Girl should any rob and mistreat them on the outside?

But by 1983 the enemy was not the middlemen on the outside, not the hard nosed who sold to private crooks for cash. It was found to be us, the membership, who complacently had allowed those on the inside of Valley Girl, the hundreds of suited operatives, to destroy the work and ideals of anonymous men far better than we but now mostly dead. All during the inflation of the 1970s the Valley Girl octopus expanded all for us, buying candy companies, nut companies, merging with fig and walnut cooperatives, acquiring distilleries and dehydrators, trolling for processing plants across the border in Mexico. In slick illustrated brochures a smiling $300,000-a-year Larry Black, white plastic hard hat perched on his enormous-jowled head, assured farmers that the new aggregate tentacles of their once-lowly raisin plant now earned a revered place among the Fortune 500. All for us, he said. Indeed, under Black, Valley Girl's novel Raisin Political Action Committee was a player in Republican political circles (which ironically had voiced no sympathy for such Democratic-spawned communitarian cooperatives); its trademark itself was well worth millions. Frightened, dull little men on tractors now blustered to the hard nosed that they too had hired their own suits, they too now were every bit as slick and sophisticated as any would-be middleman or processor. Had not they and their crafty board stolen Larry Black away from the American Standard Truss and Joint Company, replete with jet, helicopter, and luxury home? Had not Larry Black fearlessly outgunned the private locals and the multinationals alike? Was not a prominent politician to visit Valley Girl, to meet in executive session with Larry Black, to sit atop some raisin bins, straw in his mouth, and declare that he was one with its members?

The story of the fall of the Girl is an old one with predictable consequences for any who watched the evolution of a once-noble idea in the last decade. At each juncture more and more divisions were created, more offices, secretaries, more salesmen to run the now-enormous Valley Girl marketing enterprise. No longer was it a parochial Great Depression relic, founded on boring principles of fair dealing and self-help to maximize sales returns for its beleaguered, unimaginative, and profitless growers. Indeed, Valley Girl herself gave way and quite improperly, without a vote of its membership, became a subsidiary to a new "Valley Maiden," the enormous umbrella organization that had swallowed private companies and cooperatives together. Generous union contracts were given its plant workers, its administration signing on to incentive clauses, insurance policies, deferred income, cars, and expense accounts. Annual Valley Maiden reports no longer agonized how to return the largest feasible returns to its beaten-down farmers but now boasted that they might share collective pride in owning a maze of distant (but profitless and leveraged) companies and offices.

Soon (how soon only the redoubtable Rollin Buckler could decipher) Valley Maiden's returns were no longer competitive with those of private, for-profit raisin packers. The Maiden was not a private corporation or a family who grew rich facilitating the trip from the raisin tray to the grocery store. Remember, there were no outside stockholders or relatives that demanded large profits from the lucrative middleman enterprise of packaging raisins and selling them to stores and breakfast companies.

No, all Valley Maiden needed to do was to run its expansive paid-for plant at cost. All the profits of packing and selling would be divided up among its members, the members who had built the plant, worked the plant, and hired its help, whose fathers and mothers had scrimped and saved and so years ago paid off the expense of their enormous and impressive raisin factory. These anonymous ants traditionally had been the backbone of the raisin industry. They were the bane of the private packers, whose own supplies rested solely on their ability to pay prices commensurate with those given Valley Maiden farmers and so pry away enough growers to guarantee a sufficient source. Usually the raisin industry was only as strong as Valley Girl itself.

So the cooperative originally had been designed to add packing profit to the purchase price of the grower's raisins, the icing on the cake, as the lowly grower of the Great Depression was rewarded as both producer and marketer. This meager, unexciting task of stemming, washing, boxing, and selling raisins at a tidy profit the enormous Valley Maiden could not or would not any longer do. Indeed, Larry Black, confined in the southern San Joaquin, moved up north to new corporate headquarters to be closer to the culture and ambiance of San Francisco Bay, where deals, mergers, and acquisitions could be monitored and commanded in their appropriate surroundings.

Gradually the new Valley Maiden seemed to turn on its own raisin growers and become far more voracious than any private packer that profits from its anonymous and despised producers, for Valley Maiden owned its growers, had their life savings tied up in its vault as "capital retain" used to finance repair and maintenance. But now Valley Maiden and its stored hoard of membership money seemed to exist solely for its apparat, its underworked executives and overpaid workers, whose numbers grew as their productivity eroded. To camouflage its losses, to ensure their own farmers were still compensated above the private packer's clientele, to do all this, Valley Maiden began without the growers' knowledge to borrow mightily what it could not earn to pay its membership for their raisins. Only through these necessary bank drafts and a commensurate accounting labyrinth was the membership left ignorant of the extent of Valley Maiden's huge overhead and relative failure. We all kept sending our raisins to Valley Maiden because they paid us about what the private packers gave; we grumbled about this unimpressive performance, but we stayed because we remembered our ancestors had created and worshiped the Valley Girl Raisin Cooperative. But we had no idea that even our unimpressive recompense was possible only through the use of borrowed and unsecured money: We had not even earned the mediocre prices we received.

The entire shell game was exposed at the end of 1983, year zero. Then Valley Maiden could borrow no more as the once-fawning banks began to lose faith. Then its members finally were told that although their raisins sold retail for the highest prices in the industry, although they alone were to share in the profits of that marketing, they alone were to become the lowest-paid raisin growers in the valley, the lowest-paid growers in nearly fifteen years. Then any farmer in 1983 was far better off selling his raisins up front for cash to an up-front crook, not sending his harvests into the layered coils of the cooperative, which might after two or three years spit back only 80 percent of their current market worth. After deductions and retains to pay for past losses, you had to pay extra for the privilege of Larry Black and his impressive, multicolored year-end reports.

Such hubris incensed of all people lowly Rollin Buckler. Why, Rollin Buckler wished to know, did Valley Maiden raisins command the top prices in the store and the lowest in the field? And why did Valley Maiden, which was to have no middlemen, have the largest middle in the industry?

To stay afloat, Valley Maiden canceled its obligations to repay the capital moneys of its farming members. The money was long gone. I mean simply that it renounced its debt to its own members, the lifetime retirement funds of hundreds of families, borrowed for plant construction, upkeep, and modernization. It would no longer repay retained capital moneys to its own farmers. Indeed, the cooperative, in the spirit of all authoritarian enterprises that betray their constitutions, now invented a new word. All the losses were now dubbed "overpayments." Member growers, the stern board lectured, as a group had not actually "earned" from their sales the previous crop payouts that had been returned in the late seventies and early eighties. The loyal now reasoned, had not we (it was "we" once more, as in the depression), as owners of the cooperative, simply borrowed and "overpaid" ourselves? Should we not confess that we had given ourselves borrowed money that we had not earned?

Yes, they admitted at cooperative meetings, there was probably some action by somebody to ensure that we did not go elsewhere when the cooperative returns hit rock bottom. Yes, years ago the cooperative should *not* have borrowed money to cloak its ongoing failure, should *not* have paid its own members the going rate for their harvests when it had not even earned such meager recompense for them, should have learned why it cost them far more than the industry rate to process and sell raisins. Yes, the membership had been coaxed into staying with the cooperative at a time when it might have done much better with the private packers. Yes, yes, yes. We should examine where all the millions went, but only within appropriate and discreet avenues of investigation.

Still, the money we thought we had was now gone; the cooperative was reeling. Was it not better to forget the past and forgive the board who had also allegedly been victimized by their own managers? Indeed, the board now blackened Larry Black, portrayed him as the evil pied piper who had led them over the abyss with grandiose schemes and exotic tales of lucre. He was now the stuff of vilification and slander, Cicero's Verres caught in the shameful scheming of food on Sicily. Black, in turn, vociferously denied any wrongdoing in connection with Valley Maiden and claimed to have been misled by its managers.

Although in the end the various lawsuits surrounding the Valley Maiden financial mishaps were settled, so that no one was ever found responsible, those at the cooperative meetings clearly focused on Larry Black. Never liked him at all, shy Danny Dungren proclaimed to quiet applause at the solemn year-end meeting, and added that the board might not pay out Mr. Black's golden parachute after all. Once Dungren's finger was pulled out of the dike, the private cooperative gathering proceeded to voice denunciations of sorts of Larry Black himself and the assorted cars, planes, houses, and trips of the ungrateful chief executive officer of Valley Maiden Growers, Incorporated. So it was with all these dim-witted and dense that the idiosyncratic Rollin Buckler, of proud third-generation Valley Girl stock, with cousins and uncles in good cooperative standing, began his arduous struggle to inform, to educate, to organize, to reform, and to punish.

I first met Rollin Buckler as everyone else met Rollin Buckler, through his bizarre "fact sheet" mailed to all cooperative members. Who was this provocateur who sent out garish, *National Inquirer*–like headlines? His releases read in bold black letters: "LOSSES: the $27,347,000 doesn't represent our losses. That is the amount that was misappropriated to cover our real losses. WE MUST FILE SUIT. NO ONE ELSE WILL." A page of carefully reasoned arguments followed, outlining allegations that the board of directors had knowingly allowed Black to misappropriate funds, how the cooperative's failure had contributed to the 1983 raisin crash itself, how the accounting and auditing firms had either been incompetent or somehow complicit in the debacle, how the evolution from Valley Girl to Valley Maiden was designed to reward a vast cadre of nonfarming bureacrats. We had *not* lost twenty-seven million at all. More like three and four times that number over not one but several years. We should have made a tidy profit in processing and selling raisins—as the private packers did—which might have now offset the actual losses of farming our vineyards. Instead we gave our raisins cheaply to ourselves and sold them dearly to the stores so that all in Valley Maiden but ourselves would profit.

Rollin Buckler's allegations went on further. The money Valley Maiden received from the cereal companies and food stores for its overpriced raisins was enormous; yet the moneys passed on to its own growers were well below the industry benchmark. Where did those profits go?, he wondered. Still the huge plant and its overseers—glutted but not sated—had taken more, had gnawed on its own litter, the very capital retain funds of its own members. It was not enough that the farmers' middleman's cut was supposedly taken, even their banked capital was tapped as well, as each year beneath the hype and ads Valley Maiden raisin managers lost millions of dollars. Mr. Buckler had published to all that the emperor had no clothes; that behind the pious board, the massive plant, the reassuring press releases, Valley Maiden, fashioned and molded from everyone's beloved Valley Girl, was a whore. Her smile was a looter of men's farms; her salesmen and glossy mailings far more dangerous than the thief who jimmies the barn door or the vandal who shoots out the tractor's tires. They, after all, stole and destroyed only commensurate with their limited ambition and intelligence. And, Mr. Rollin Buckler concluded, he had the facts, documentation, arithmetic, and proof to expose the entire sordid story. And, Mr. Buckler announced, he planned, alone if need be, to do something about it!

The effect of Buckler's letter was electric; suddenly the mist of Larry Black's decade-long fog cleared for member and nonmember alike. Immediately, within hours. The once bleary-eyed farmers now woke up and cursed the board, Larry Black, the raisin industry, and wondered how they were to buy their diesel fuel, maintain their living expenses, begin their pruning, or meet their land bank payment when their raisin returns were to be scarcely a third of last year's and their thousands of accumulated capital retain wiped out.

How, they growled, could a man get $1300 to $1400 a ton in 1982 and then a year later receive $450 for the same two thousand pounds of raisins? How could a hundred-acre grower have $50,000 in retained capital owed to him in December by Valley Maiden and have it all gone in January? To the cooperative members it became the Great Raisin Crash of 1983 *and* the Great Cooperative Crash that ensured them they would lose an additional $100 or $200 a ton over their near-broke brethren who sold to the private packers and took their paltry cash up front.

Three hundred of them, red eyed and foaming, not at all like the politely indignant at recent private cooperative meetings, crowded into a local rural school to hear Rollin Buckler, this rotund Gracchus back from the dead, this Cincinnatus as field general, go beyond his written statements and in person uncover the ruse. Surely Rollin Buckler would tell them where their money went, how they were to get it back, and what they were to do in the future.

In his earnest, dry, and methodical voice, glasses down on his nose, his enormous briefcase and stack of documents at his side, he went on for two hours, documenting the farmers' alleged losses and the extent of the supposed Valley Maiden losses. Like all mobs, the crowd grew heated, both demanding that he march out immediately to sue on their behalf, alternatively threatening Buckler that this was a private, in-house affair, that this horrendous (and nearly unbelievable) tale should *not* be aired. Indeed, they forced outside the hall the local media who sniffed some type of populist uprising and worried about a petite *60 Minutes* exposé.

After three hours, two of us farmers found ourselves elected as part of Buckler's Raisin Action Committee. For the next five years, I was carried along on the coattails of Rollin Buckler's personal crusade, involving thousands of dollars in legal expenses, hordes of Bay Area and Los Angeles attorneys, threats, court appearances, and all sorts of base attempts to acquire and then ensure his silence. In the end it was quite a legal circus, this effort to untangle the Gordian knot of Larry Black's conglomerations. Other lesser, copycat groups filed suit; Larry Black himself filed suit; his deposed underlings filed suit; Valley Girl filed suit; Valley Maiden filed suit; the auditors and accounting firms filed suit; everyone quoting, misquoting, using, and abusing the hours of accounting and legal work of one Rollin Buckler, who alone apparently knew and could document how Valley Maiden had managed to borrow in order to pay its members what it could not earn.

The local and lesser valley legal establishment, the simulacra that sniffed and shuffled around the courthouse, hearing rumors of accounting liability and getting a whiff of multimillion dollar insurance policies, entered the frenzy with tusks lowered and tongues extended, desperately rooting out victims and plaintiffs. Ninety-year-old women and senile octogenarians were gored by probate barristers and traffic lawyers to serve as aggrieved victims of Larry Black and his now odious Valley Maiden conglomerate. Most lawyers, absolutely ignorant of Valley Maiden protocol in general and the raisin industry in particular, seemed simply to translate or retype Rollin Buckler's fact sheets and exegeses into hastily drawn-up legal briefs in preparation for more formal suits. It was miraculous to see Rollin Buckler's prose contained on legal paper generated for $300 an hour fees. A few of the Fresno legal mongrels even managed to gnaw into a rarefied Bay Area settlement conference or two and obtain from the more urbane and pedigreed hounds a $250,000 bone. Raisin growers had no money; Valley Maiden had no money; even Larry Black had no money; so they focused on the deep pockets of the accounting firms.

And Rollin Buckler? He suspected that some auditors might be at fault, that their apparent oversight meant financial liability (the auditors denied any improper accounting, claiming that the audits they were hired for didn't include the key records involved). But in the end he turned out to be oblivious entirely to the money to be made. Absolutely uninterested in cash he was. I saw him irritated only with the more brazen who looted his work verbatim and then arrogantly quoted it back to him, assured that packaged with forensic gesture and bellowing voice, Buckler's theses and research might perhaps take on the shape of careful legal scholarship and brilliant judicial deduction.

For a time Rollin Buckler was even quoted on local television, in the *Fresno Bee* and surrounding papers, on the tongues of the valley's hundreds of raisin growers, as he sought desperately to reclaim ten, twenty, perhaps even thirty cents on the dollar of the retain funds for the membership. He mailed out briefs, letters, gave interviews, attended meetings, all free, without recompense, without swagger or brag.

In fear of this funny man who stuck his nose in reports and audits each evening until the early morning, Valley Maiden announced sweeping changes, shrank away, and from up north spat back out the old, familiar Valley Girl. Salaries were frozen, bonuses curbed, overhead cut as the Girl struggled to survive even as its more independent, more vocal growers left in droves to join the hard nosed.

As the cooperative remnants damned and dreaded Rollin Buckler, they became more like frugal, upright Rollin Buckler, and so they miraculously survived, abandoning all the practices that he had identified and then condemned. In a sense they were saved by Rollin Buckler of all people, who, now reviled by all, retreated in dejection back into the anonymity of his trailer house amid hundreds of Thompson seedless vines.

Although Rollin Buckler lives a mere four miles away, I have not seen him in five years. My last memory of him was at a tense hearing in the Fresno County courthouse. His Raisin Action Committee had retained on behalf of its growers a very suave, utterly dishonest plaintiff's attorney from San Francisco. His prominent name supposedly had given us a sort of respectability and legitimacy to battle the larger Los Angeles firms of the antagonists. In the settlement conference, with the accounting firms, Valley Girl, Valley Maiden, Larry Black, other growers, and legal groups, this lawyer, our majestic attorney, Les Hightower, of Reynolds, Hightower, and McGowan fame, came out beaming that he had received in this particular round eleven cents on each of our lost retained dollars. The farmers of the Raisin Action

Committee, he grinned, due to his rare negotiating skills, had now lost but eighty-nine cents of each of their once pilfered dollars. His own cut, agreed on in secretive meetings among all the hostile lawyers the night before at the airport in San Francisco? It was pried out by us that it was a little over $250,000. I thought of Juvenal's "who will police the police?"

This staggering sum was in addition to the $70,000 we broke raisin growers had paid him upfront. Les, Rollin Buckler inquired, surely had already received enough attorney fees through his considerable ongoing (and inflated) payments from Rollin Buckler's Raisin Action Committee? Could he not at least reimburse these growers their $70,000 or perhaps have the thirty or so lawyers agree to subtract that sum from their addition revenues? $250,000 in profits was excessive; $250,000 taken by Lester Hightower from raisin growers on top of a $70,000 up-front fee seemed criminal.

Especially was this both honorable and fitting since Rollin Buckler had convinced two hundred broke farmers to hire Mr. Lester Hightower and associates in the first place; these men and women as of yet had received not a dime for their lost millions and had not foreseen that most of their lost raisin money might be rerouted from insurance companies to lawyers. Rollin Buckler himself had done all the research for Reynolds, Hightower, and McGowan without compensation. Indeed, I piped up that he, Rollin Buckler, had just about written the briefs himself, coached silver-haired, Italian-shoed Lester Hightower until seconds before his court appearances, and driven up to San Francisco weekly when this far more obtuse, duller man could not understand a point of law or a wrinkle in the raisin business.

Absolutely not, Hightower chuckled at the naïve. Rollin Buckler was to be congratulated for his efforts. Was the noble Buckler, Lester Hightower declared, not a pioneer, a Joan of Arc? Was not he alone responsible for winning a few tough cents back for the raisin men? Did not this agrarian knight frighten back Valley Girl from Valley Maiden through his selfless and heroic efforts, all ending in needed reform? Mr. Rollin Buckler was a noble crusader who, in the tradition of all courageous Samaritans in a cruel and unforgiving world, should be satisfied only with the accomplishment of good deeds for his fellow man.

Yes, Mr. Hightower continued, he himself had received $70,000, which quite generously had covered *all* his hours of work, expenses, and travel, but that was understood as only a nonrefundable retainer fee. This present $250,000, this quite unexpected added recompense, was his alone as legitimate profit for squaring accounts in an equitable and legal fashion, in an ethical and sophisticated way, a proper way for the benefit of a group of otherwise largely ignorant and small-minded raisin growers.

Before smug Lester Hightower could return to the settlement room to sign his papers and take his money back to San Francisco, before he could finish his peroration on his concern for the snookered farmers, and before he could remark on his pleasant visits to Fresno, red-faced Rollin Buckler interrupted. No, Mr. Hightower could not return to the conference and take any more unearned money from two hundred broke raisin growers. Did we, Rollin Buckler turned and asked me, any longer need a Lester Hightower? Lester Hightower had indeed been paid his (excessive) fee of $70,000. Could we not ourselves as agents in good faith sign the papers and turn the $250,000 in legal bonuses back to the farmers? Barring that, could not a lawyer cousin or friend of Rollin Buckler now drive up from Alma, drive up right now to represent us and assume power of attorney in the signing process for a fee of perhaps $200 to $300? Could not then Mr. Hightower and associates be terminated as of now and not return to the settlement room? Might not an Alma probate attorney take his place from here on out for 1/1000 of Mr. Hightower's price?

Mr. Lester Hightower returned to the settlement conference and came out again with money to pay back the farmers' $70,000 that they had paid him. We two hundred were refunded our $70,000. The hordes of attorneys, Lester Hightower included, walked out with $250,000 each, all the Los Angeles and San Francisco species, mumbling that they had missed the last flight out of Fresno and inquiring where, if anywhere, the fine restaurants and exclusive bars were now to be found. It was as simple as that; a one- or two-minute request, a few shouts from inside the room, and the 200 growers' up-front legal fees were returned. Perhaps all the other thirty attorneys cut off about $2,500 of their loot to help out poor Mr. Hightower. Why not chip in when their individual take was a hundred times that sum? And the growers' lost retains? Last I checked, we had been paid back about eighteen cents on the lost capital dollar and, of course, nothing for the many years in which the cooperative's yearly returns were lower than the private packers'.

If you loot in this country and still wish to save your retirement and good name, your theft must not be in the thousands, not the hundreds of thousands, but in the many millions of millions. To this day, each time I see the evening news with its nightly thuggery, clips of quick-stop holdups and car crashes into department store windows, I still think of Larry Black's tally. It will take a hundred thousand liquor store robberies from Bakersfield to Sacramento to match the score of Chief Operating Officer Black, Mr. T. Larry Black, whose disdain for hundreds of small raisin growers earned him a six-figure retirement. When he was fired, I recall he announced something like, "I worked my fool head off for those raisin growers and now I am going to have to work my fool head off making them pay me my retirement." And so he did.

Two years later, after our meeting in Fresno County courthouse, Lester Hightower, our lawyer, a dean of the Trial Lawyers Association, Pacific Heights denizen, athletic, sixtyish, brushed gray hair on his temples, debonair, social butterfly, and born liar, dropped dead. Passed on prematurely did he, with his Bentley, shimmering suits and soft shoes, enormous home, pretty family and girlfriends, without even a blink or a nod from his former clients, Rollin Buckler's raisin men down south of Fresno. I heard that after weeks of silence from Hightower, Rollin had finally in exasperation called the law office. "He's been dead a week," the bothered and passionless secretary announced and hung up. "He dead," right out of Conrad. Supposedly the mayor of San Francisco attended Mr. Lester Hightower's funeral.

After all, Hightower had in his glass office a huge watercolor painting of his own sprawling Pacific Heights estate; after all, he had his bushy eyebrows washed each week, razor-trimmed, and blown dry.

It is not that farmers dislike instinctively all the accoutrements of late twentieth-century urban success, the impressive but largely unnecessary stock in Mr. Hightower's sizable corral. We are not naïve, unsophisticated whiners; we are not the envious who pine over the pouch of the rich. But if one farms, like one Rollin Buckler, keeps to himself, finds ample shelter, sustenance, and security in growing food, in working with his hands and mind to decipher the secrets of fruit production, he will, I promise, in no more than three or four years' time become baffled by $50,000 cars that drive no faster than his pickup, by multibedroomed homes where sleep no more than in his own, and by frequent lavish meals where the food is less tasty than his own daily fare. Why, to lasso in all that, the farmer would have to triple-crop his soil, wear out his ground, drive his tractor ragged to meet the tab! The poet Menander surely could not have had all that in mind when he concluded simply, "A farm is the best gift a man could have."

What about the other lesser relish of the plutocratic Mr. Hightower? The successful urban male's expensive gold rings and chains, the cellular phones, the Italian loafers, Italian bicycles, spandex, and Rolex watches, for what purpose are they? A hard nose might hog the ditch for a tractor or haggle over the rent of raisin bins for a new set of disk blades or hydraulic hoses. But to lie and connive, to loot and extort for leather, metal, and plastic for the foot, finger, or cuff? For not much more than that were Mr. Hightower's whine over fees and his enormous nose for profit. All that was about as far from producing food with muscle and horsepower as your child's tinker-toy creation or Monopoly gameboard.

I wanted to believe that Rollin Buckler had in part caused the death of Lester Hightower. His years of correcting Lester Hightower, of watching Lester Hightower, of *teaching* Lester Hightower about the law and the ethical conduct of law was lethal to a count of fluff and braggadocio who sat in a glass office, doors and walls all glass, high atop the glass of San Francisco, not in a sweltering mobile home north of Alma.

When you are better dressed, more impressive visually, with mellifluous voice, and confident gait, cash and bond laden, how might an impoverished hayseed from Alma, California, know more but fear less, want less but need more, give more but receive less? Shortly before his unexpected demise, confronted with this farming oddity, this agrarian exemplar on the phone and in the flesh, Mr. Hightower himself said that he no longer enjoyed either the San Francisco visits of Mr. Rollin Buckler or his own rarer flights into Fresno.

I can confirm that. When he walked into the luggage depot at the Fresno airport, he at once darted over to a phone. There I heard him tell his secretary back on California Street in San Francisco that he would be out of this God-forsaken place on the six o'clock flight the next morning. When he emerged from the phone stall, Rollin Buckler was in his face, methodically reeling off Valley Girl profits and losses from 1975 to 1982, quoting an obscure case law about cooperative forfeiture, lecturing about the strengths and weakness of a local Fresno judge, adding as condiment a poignant anecdote about a broke raisin grower who had written he could no longer contribute for Mr. Hightower's spiraling fees. Mr. Hightower had not yet reached the baggage claim at the Fresno airport and already he was not looking forward to his reservation at the nearby Holiday Inn, a Raisin Action Committee splurge that Rollin Buckler hoped would be deeply appreciated in the months to come.

Lethal even in small doses to one of Mr. Lester Hightower's ilk was this strange paladin with borrowed truck, trailer house, and black plastic glasses, this Plutarchean refugee from a vanquished world of good and evil, villain and hero. He alone had done all Mr. Hightower's work. He had explained the entire complicated cooperative fiasco to any lawyer and accountant who would listen and absorb and profit. He had taken not a cent for his *thousands* of hours of work, not a salaried job, not a 10 percent commission, so worried was he about winning twenty-five cents on the lost dollars for the defrauded membership of Valley Girl Raisin Cooperative, most of whom now hated him for exposing their own naïveté and imbecility. Everything Lester Hightower wrote, every brief, interrogatory, and appeal; everything he said, every court speech, lecture, and private barter, came from the mind of Rollin Buckler. Mr. Hightower, of national legal reputation, fluted pillar of Bay Area society, was exposed to his clients and peers as a brainless puppet. Like some lewd mime on the old Roman stage, he could only add gesticulations and a few empty asides to the convincing script of Rollin Buckler.

I do not think that Rollin Buckler in this lifetime will yet again emerge far from his trailer house. He will not leave his sandy vineyard to joust once more with the conquering magnates of California agribusiness and their sordid hirelings. For even in defeat Rollin Buckler had made his point. He has told us what we always knew and should have known, what the old prophets told us between our dancing bouts with the golden calf, what lies locked in the forgotten books on the dusty shelf, in the scattered atoms beneath the gravestones, proved to himself at least that it is still even now not a man's title or salary, not a man's clothes or looks that radiate a power, a way to see—and then to do—the good. Much less does education or repute ensure character. Indeed, in America now the brush with the university or government can more often dilute and pollute what little good is inside to begin with. Remember that the next time a silver-haired bond trader or tweed-suited university president is lauded by his fawning aficionados for a lifetime of "public service," keep that in mind when Mr. Degree's vocabulary is deep, his heart shallow.

"I'll get right on it," "let me do all I can," and "I'm the man for that assignment," however confidently spoken, however brashly barked out to superiors and comrades alike in the workplace, mean little to mute species of plums and grapes. You do not write reports to the vines why you wish to work with them. You do not submit a *curriculum vitae* to prove that you are their match. There are no *res gestae* to be bandied about to demonstrate worth, for there is no worth if you cannot produce food for a profit in the here and now, in the long year of frost, hail, drought, rain, equipment failure, insect attack, broker assault, and simple barrenness that has no cause. Character is more likely to be fashioned by a funny-looking man on a tractor, who must learn the ways of vines which he cannot fool, cajole, or impress. His simplicity and honesty are sharpened, tempered, and held straight by the unforgiving arts of irrigation, cultivation, and harvesting, he who is shamed by nature and so can shame those powerful in his midst when his rope is snapped and he enters their fray.

IV. Mere Renters

And what of us? Is there at least place for us in the agrarian pantheon, we who worked with and beside Misters Kobashi, Ponce, Buckler, and the dozen others in the acres of renown whom I have not described; we who were far better men than the late Kapoor, Kaldarian, and Dungren?

The answer is, I am afraid, no. The reason for such a harsh judgment on us, the vainglorious, the equity gobblers, and money grabbers, remains controversial and sensitive, and it cannot be fully disclosed, at least not now without hurt to family and self especially. (A job or two in town or a queer virus, I see now, is not really an excuse to exit the vineyard.)

Surely, then, one can at least *rent* a space in the pantheon for a season on the strength of a single great notion or deed? Did we not ourselves have a meager brush with excellence before our fall, before the Regal Reds, and the Royal seedless, before the great crash of 1983, when my uncle had not yet blasted his heart apart, when the tumor in my mother's brain was still small, if not nonexistent, when water was in Louisa Anna Burton's pond and the land around it still planted in vines, when there was some money in the bank and the housing developments were still quite a distance away, when this family was not scattered but lived in health and worked so very hard as one?

We did. In those calmer, better times of 1982, in the penultimate harvest of September, 1 B.C., we together, every one of us, brothers, cousins, mother, and father, saved our raisin crop and hence our century-old raisin farm in the rain of but a single day, with the help of bad-backed Ulysses Ponce and the magnanimity of his highness, ChiChi Martinez, and with the unexpected twilight arrival of Arturo Cazares and his De Soto wagon full of raisin-tray-rolling kin, with the clan of Javier Mota puttering in not far behind. From that great endeavor of salvaging two hundred tons of raisins when others could not, from the last big check from the exhausted coffers of the Valley Girl Raisin Cooperative, we built sheds and houses, purchased tractors, vine cutters, and manure, and brought in a frenzy of strange, hardworking men eager to take our hard-won raisin money before the fall. A single season of excellence, a single day's struggle in the vineyard mud, freed us, as we then imagined, from debt forever; primed instead were we for the real boom of the eighties and nineties to come.

We siblings once all lived shouting distance apart on our great-great-grandparents' land, farmed every acre of it when hustling Bus Barzagus used to visit and mentioned those new tasty apple-pears and the richness of his mountain soil, when we saw and raised him with our own beautiful newly planted Regal Reds and Royal seedless, when Larry Black was but a pompous nuisance, when Taro Kobashi surely an unknown field man for a prosperous packer, and nondescript Rollin Buckler, silent and alone, must have worked in oblivion his dad's vineyard, battered after his losses in his tiny linoleum business.

But now, as the Old Testament prophet once warned, "The harvest is past, the summer is ended, and we are not saved."

CHAPTER NINE

THE FIRST AND THE LAST

Family Farming at the Millennium

I.

There is no controversy over the status of agriculture in America. By any yardstick one uses, we are now in the penultimate stage of the death of agrarianism, the idea that farmland of roughly like size and nature should be worked by individual families. Do percentages about farm income and property ownership reveal the real story? If so, nearly 1 percent of farmers now account for the majority of farm income, while 88 percent of agrarians earn less than $20,000. By the year 2000 a mere fifty-thousand farms—the number of small operations that once surrounded the typical medium-sized American city—will produce 66 percent of all the food in America. The parity figure—the traditional indicator of real farm income—in 1982 was lower than it was in 1932, in the midst of the Great Depression. Five percent of American landowners own 75 percent of our land, the bottom 78 percent own just 3 percent. In states such as Hawaii and Florida, 5 percent of agricultural concerns own 90 percent of the land; in California one twentieth possess over 80 percent of all the farmland in production. There has been no similar revolution in America—not in the universities, not in corporations, not in government.

Perhaps absolute numbers about the rate and nature of agrarian decline paint a more accurate picture. Two thousand family farms a week vanished in the decade of the 1980s; two thousand acres of their land were lost to urbanization *each day*. From 33 percent of the population in 1910 to 25 percent in 1935 to 10 percent in 1955 to 5 percent in 1967, farmers today are less than 2 percent of the American populace. Most feel that the figure is inflated by weekend part-timers whose prime income is not agriculture—the number of true agrarians may really be somewhere between 0.5 percent and 1 percent of America. Thirty-three percent of commercial family farms (with annual sales between $50,000 and $500,000) face financial uncertainty; 31 percent had accumulated debt worth 40 percent of their assets. Their fate is sealed.

There is little need to go further with such citation since all studies by both conservatives and liberals alike point the same way and can be reinforced by personal anecdote and direct observation. Surrounding our forty acres of trees and vines in Lundburg where we once planted the Regals and Royals, every farm but two has now failed. But *all* the land there continues in tree and vine production under alternative management schemes. One forty-acre walnut orchard is kept in a family, since both the father and son have full-time jobs with the power company and railroad; they can subsidize the grove's losses as the price of their picturesque abode. It is a beautiful thing to live inside a walnut orchard in the valley summer. Similarly, our eighty-year-old Swedish cousin, a retired schoolteacher, continues to lease out his adjoining nectarines and peaches to a corporation; his small piece in Knut Hendrickson's once-considerable nineteenth-century farm has likewise not been sold.

But the much larger foreclosed tree fruit orchards across the street are now parts of Elton Wood Enterprises, a local manager for unknown corporate investors, which buys up orchards and vineyards to integrate them within a larger packing, shipping, and brokerage conglomerate, the Wood Group. Usually, tottering agrarians are contacted by Elton Wood, whose line of credit is as unending as it is mysterious. The less desperate are equally welcomed into the group through long-term lease arrangements, whose rental payments have often proved far more lucrative to the farmers than their own previous failed efforts. Our cousin, for example, receives a tidy profit on land the corporation leases and loses on; they can pay his rent from their profits in packing, shipping, brokerage, but not the growing, of his fruit. Last year he took in $27,000 in rent for his forty; on our identical adjacent parcel we lost $8000 for the year's toil. I think in a few years or so the Elton Wood Group will own all of our land, both of the original Hendrickson forties alike.

The other yeomen sell out to Anthony "Tony Boy" Locatelli, Inc., a self-made multimillionaire who started with forty acres in the depression, bought foreclosed farms and undeveloped ground, and so expanded up to ten-thousand acres of trees and vines with an enormous supporting infrastructure of packing and distribution—a family corporation whose shiny vermilion trucks, vermilion pumps, vermilion ladders, vermilion houses, and vermilion farm equipment are a local fixture. His vermilion home is atop a mountain, said to be equipped with an elevator drilled through the granite rock, and known for its unrivaled vista over his vast domain of nearly a million fruit trees.

I suppose that Ron McCullum, who, as I mentioned in chapter 1, is the president of three local land-development corporations and real estate operations, was fronting for either one of the latter two megafarms. He said our Lundburg ground was quite lucky to be nestled between the land of Wood and Locatelli. In his calculation, my great-grandfather Knut Hendrickson's legacy would draw 20 to 30 percent more than it was worth as farmland per se—once the two agribusiness titans began their inevitable joust over another small tessera for their respective mosaics.

But in his market appraisal of our acreage, Mr. McCullum was not entirely critical of the Alma land of Rhys Burton, the tule pond homestead of Louisa Anna Burton and her son Willem. Not at all did he dismiss its salability outright, despite the old vines, odd configuration, muddled ownership, and the continuing raisin depression. (He did not see the south side.) The expanding Alma subdivisions and the new Wal-Mart are now less than two miles away and thus very good omens. Alma may no longer be a tiny agrarian *polis* of like-minded peers, may be instead an ugly all-night edge city on California's main freeway, but whatever it is, people want it, and it is growing toward us. In his professional tone, with the air of knowing the big picture in store for us all in the valley, McCullum went on: "Your land here near Alma, to be frank, is not as prime as the forty in Lundburg, so corporate interest won't be optimal. But look, urbanization and the freeway, even if the sprawl is from the wrong side of town, is still a positive development, and that will maximize your farm's value, at least enough to ensure real profitability at sale. So I think we could still move aggressively for you on both units."

Whatever the difference in location or soil, the actual farming predicament around our home place in Alma is exactly the same as in Lundburg; both parcels are, after all, operating family farms and therefore fated. So I do not know which is more lethal: the spectacular yearly collapses of our peaches and plums in Lundburg or the dreary and predictable unprofitability of Alma's raisins. Still, to be fair to the raisin depression, in that one- or two-mile radius around the home place, once again all but two farms either are rented out or survive through off-farm income of both husband and wife.

The most perilous family farms seem to be those in our own size range, between eighty and two hundred acres. They are too large to manage on the weekends, between schoolteaching or a construction job, and far too small to be integrated vertically within a chain of distribution and processing appendages. Their need for capital precludes self-financing and yet does not require a line of credit in the million-dollar spectrum; such perilous exposure at least forces the bank to show deference lest you take some of them down with you.

On farms like ours, everything from repair to replacement planting to the activity of the workers themselves is slowly grinding to a halt as capital, not labor, vanishes; we are like plodding, sick stegosauri when the temperature falls below freezing. Long gone are cash and equity that might stanch any future hemorrhage, the type of catastrophes we and others once incurred with the raisins, the Regals, the Royals, their succeeding failed varieties, and the unexpected deductions, bills, and dumped fruit, courtesy of Messy Messingale, Ross Lee Ford, the Effin team, Valley Girl, and a host of other brokers, packers, and distributors too numerous to mention. We have depleted the last twenty years of my mother's life, her aggregate saved income from her forty-fifth to sixty-fifth year.

Yet my brother and cousin would object to "depleted"— "borrowed" is their preferred term. And in a perfect world they would be right. They argue that in this rather long cycle of agricultural depression—in the life of our farm, well into the second century—we have drawn on but not lost our capital. True, we are but a frost, an untimely rain, a Ross Lee Ford consignment sale away from oblivion, but so were Willem and Rhys Burton for much of their lives. Better, then, to see us but a season away from resurrection as well. Every real farmer still left knows the refrain: In the next three or four years when the bounty returns, the ample profits of our trees and vines will replenish the reservoir of capital, and all will be set aright. Loans will be repaid with proper interest; family sacrifices will be rewarded magnanimously; superficial cracks in clan solidarity will disappear as raisins top $1,200 a ton and more, as peaches and plums also return to their generous prices of fifteen and twenty years ago. Tragic it would be, they add, to lose two century-old family farms in the year or two before the boom of the late 1990s has begun. And so they search desperately in the meantime for the modern equivalent of Rhys Burton's Gower nectarines, the rare fruit of the resurrection that might yet save an entire clan.

But I have lost hope and so write now in great haste that this account might yet see light while I can still claim farm ownership. I think it wise to remember the Burtons and Hendricksons while Logan and Rhys, alone of us now, are still true farmers, while I can still walk for a year or so longer on the south side to record the memories of Willem, Orin, Rhys, and Catherine without asking the permission to do so from a foreman of Ross Lee Ford or Elton Wood.

But what is the more abstract response of the learned and educated to this extinction of the American yeoman, to the end of the Rollin Bucklers, Bus Barzaguses, and Taro Kobashis, to the destruction of Alma and Lundberg as small towns of skeptical homesteaders and yeomen? There are, I think, two schools of thought: the self-proclaimed nurturers versus the exploiters, the agrarians against the agribusiness men, the family farmers opposite the corporations. Books abound with titles such as *Saving the Family Farm* and *Ethics in Agriculture*, where data and argumentation are eloquently put forward against the present trend, where the moral high ground is claimed for the preservation of family-owned farms of moderate size.

But agricultural economists, biologists, and most economists counter that the move toward latifundia is inevitable. Based on proven economies of scale, they say farm corporatization is as American as the rise of the industrial state in the nineteenth century. As small car companies gave way in the early twentieth century to Ford, GM, and Chrysler, as oddball tinkerers and nuts like Studebaker and Tucker were absorbed by their more extensive brethren so that we in America might enjoy at last air bags, antilock brakes, and power windows, so too agriculture will be and must be transformed. As cars are made from ores and oils, so too agriculture—which likewise is of the soil—has no exalted status that might exempt it from the unyielding American canons of capital, finance, and production. The more bold of this now-victorious school take a perverse delight in the destruction of the yeoman and scoff at the admittedly romantic arguments to save the family farm. They sniff beneath the bib overalls and denim veneer of their opponents and think they detect the unmistakable musk of the university agitator, the America Laster, and the ubiquitous social provocateur. More reasonably, though, they simply, and with some cause, see the American farmer as a relic, a bothersome vestige of a time thankfully past and better left forgotten, a rebuke to themselves and their world. I think half my brother's problems with the bank could be finessed with polyester and polo shirt, leather satchel, laptop, and spread sheets—with cigarette, handwritten charts and graphs, binder-paper figures, and mud-covered boots put in storage when he walks through that awful tinted-glass door.

True, the adversaries of agribusiness, the advocates for agrarianism, are an assorted bunch. We few left in farming mostly despise those who are dedicated to saving us. Some are the radical environmentalists and utopians at the fringes of every reform school, in addition to a few conservative farmers suddenly converted to the cause of property rights, income tax evasion, and assault weapons once the foreclosure notice arrives in the mail. But the majority who fight the corporations are an idealistic but often rather protected species of conservationist and renegade university professor.

And thus the two sides battle, left and right, idealist and realist, communalist and capitalist, for the soul of the American farm at conference and on editorial page, before Senate committee and in journal and book. And the fight often takes bizarre twists: conservative, free-market Republicans argue for increased federal subsidies for their home-state corporations; big-government, tax-and-spend Democrats attack reclamation laws for the wealthy. "Nothing more than welfare and a piratical raid on the federal treasury," these born-again Democratic tightwads charge.

Farmers, baffled at foe and ally alike, stand in the vortex, dumfounded that conservatives like themselves now advocate radical farm policies that destroy their land and families; they are even more surprised to hear distasteful liberals champion their notion of conservation and protection from continuous government onslaught and corporate subsidy. Democrats have outlawed farmers' beloved chemicals that have poisoned their well water and made them sick and sterile; Republicans have protected the hated laissez-faire world of the broker that has robbed them. Democrats demand fair and managed trade overseas that enriches domestic farm export; Republicans champion absolutely free international commerce that has dumped subsidized foreign produce on the market and helped to ruin the American farmer. Democrats say go easy on immigration reform, thus ensuring cheap labor for the farmer's vineyard; Republicans decry the Third-Worldization of California and so seek to bar those who would alone work in the field.

But the final verdict on the future of the American farm lies no longer with the farmer, much less with the abstract thinker or even the politician, but rather with the American people themselves—and they have now passed judgment.

They no longer care where or how they get their food, as long as it is firm, fresh, and cheap. They have no interest in preventing the urbanization of their farmland as long as parks, Little League fields, and an occasional bike lane are left amid the concrete, stucco, and asphalt. They have no need of someone who they are not, who reminds them of their past and not their future. Their romanticism for the farmer is just that, an artificial and quite transient appreciation of his rough-cut visage against the horizon, the stuff of a wine commercial, cigarette ad, or impromptu rock concert.

You see, we in America have been running away from the isolation, uncertainty, boredom, toil, and drudgery of the farm for a century and more now and in our lifetime have finally escaped it entirely. No longer do we need to shovel furrows at seventy-five, play Scrabble with gramps on Saturday evening, or lose our hearing to the tractor's roar. We need not worry about rains and hail that wipe out years of labor and capital, much less coyotes which eat our kittens or brokers who devour our young. A check every Friday, a barbecue every other Saturday, Fescue or Rye the real difficult choice: these are what we want. We desire the security of the corporation and bureaucracy even as we hate what we become, and so we run from the farm only to dream that it might save us all yet.

Even our abstract thinkers, the subsidized policymaker and university professor, the musician, and the artist, ignore the farmer. In their view, the cities are the great conductors of civilization's energy, the countryside but dull insulators where the juice of creativity is shorted and grounded out. They write essays on liberty and responsibility, on the moral and ethical sense, on God and man, but all in a void of the potential, of the abstract and theoretical, without any notion that all utopias—as Aristotle and Plato knew—start with the land, that all of man's thought—as Pericles and Socrates saw—must be balanced by the physical, that the farmer and his soil—as Euripides and Aristophanes believed—grow the first good citizen.

And so even the learned—especially the learned—fail to resolve the dilemma that continues to plague us: We Americans have become ever more unhappy as we grow fatter and more free. With more leisure, more bounty and affluence, more safety, sanitation, and elegance, more spandex between us and the dirt and grease that are the cost of battling nature, can we, free of the craggy, unpleasant octogenarian, of the self-employed skeptic, still maintain a republic founded in another age?

A body of wisdom in Western civilization from the Greeks down to Jefferson says that we cannot—not without this agrarian prerequisite. It is impossible because a reservoir, even if it be tiny, of tough, unpleasant, and independent people—Attic farmers, Roman yeomen, Swiss pikemen, American farmers—is always needed in a democracy to ensure against the tyranny of an urbanized and uniform majority, to provide a pristine reminder of the original plan when others have forgotten or rejected it. The metropolitan, whether he be in Montezuma's city, imperial Rome, or Los Angeles, always loses a pragmatic view of nature and man's effort to master it and so ultimately finds comfort in, not unease with, the daily replication of thought, action, and comportment among like kind, comfort in a bland sameness and stasis which finally is so lethal to the idea of participatory democracy. And in America democracy and capitalism—the sociopolitical system which alone works—will ultimately bury us all beneath a sea of affluence and liberality if there be not a body of independent thought and action outside our reach. And that counterpoint and corrective to the evolution of American culture, as we have learned the last decade, is not to be found in the corporation, the government, or the university.

I have argued at length elsewhere that the entire history of Greece, the origins of Western civilization itself, lies in the countryside. The Hellenic *polis* and all which that institution implies—art, literature, notions of constitutional government, free speech, private property, direct vote of the citizenry, separation between political and religious authority, the right to bear arms, and the chauvinism of a middle class—*followed* from, did not precede, a vibrant agrarianism. Athenian democracy was not a radical departure for the Greeks but rather a further and logical refinement of a prior two centuries of agrarian egalitarianism.

Once Greek farmers—the *geôrgoi* whose ancient farms we still see in the Greek countryside, who frequent the stage of Greek drama and appear in epic and lyric poetry—divided up the countryside into individual plots, erected homesteads, devised ways that allowed each man to farm on his own as he saw fit, served in the assembly, and claimed his slot as an infantrymen in the phalanx, Greek civilization followed. No wonder Aristotle said the agricultural population was the "best," the only ones who work outside the city's walls, who come into town in times of crisis to use their egalitarian wisdom to save the state from itself. In his review of the history of Greek constitutional government, he knew that the rural yeoman had created and maintained egalitarian government: "When the farmer class and the class having moderate means are in control of the government, they govern according to laws; the reason is because they have a livelihood, and they are not able to be at leisure, so that they put laws in control of the state and hold only the minimum of assemblies necessary."

Yet once that Greek system of autonomous city-states based on agrarian notions of small farming, constitutional government, and infantry militias vanished, classical Greek culture was lost. Literature became stylized and repetitive. Taxation and military expenditure soared. Authoritarianism replaced popular government. Without the agrarian infrastructure there was no middle to frame, support, and mend an egalitarian society. All that is a historical fact, not a romance, not an agrarian yarn.

We can see the trend from Greek archaeological and literary evidence. In Hellenistic and later Roman times (338 B.C.–400 A.D.) when family farms were about gone, either homesteads are reparceled and tied into larger blocks or enormous numbers of scattered farms come under the ownership of one big man. Crop raising falls into patterns of monoculture. Formerly good agricultural land is abandoned altogether. Rural settlement and population decline. In some aspects society returns to the Dark Age culture before the advent of the Greek *polis*.

True, this Hellenistic world and Rome as well (we are now constantly reminded by classical scholars devoted to binding up contemporary society's wounds) was a more multicultural, dynamic, and wealthy cosmology. At that time states, like our America, were far better adapted to capital formation, commerce, and the vast extension of military power, far more flexible with the granting of citizenship and the marshaling of large, diverse, and lethal armies.

True, there are paradoxes within agrarianism: To live and work the farm is to be set in one's way, to be blinkered and uneasy with innovation and cultural upheaval. But still, the end of agrarianism, ancient and modern, has had a price. After the fourth century B.C. there was an undeniable chaos in the Greek world. Individualism had gone mad, a product of growing political turmoil, increasing inequality, the loss of social cohesion, and enormous military expenditure among the city-states. Those were precisely the perennial problems which the blinkered world of Greek agrarianism for nearly four centuries had assiduously sought to address and which we accept as the price of material progress.

To the modern literary, philosophical, and theoretical historian of the 1980s, the despotic Hellenistic world, like our 1980s and 1990s, was an intellectual feast of increasing diversity in philosophical approaches, growing organization and sophistication of academic enterprise and inquiry. It was a thinking man's gala, a multicultural pandemonium much in vogue again, much to the taste of the most recent university denizen. But the post-*polis* in antiquity was also a world—again, like ours—roughly contemporaneous with the period in which rural life and the majority of the citizenry were rapidly vanishing as independent and powerful forces for political, social, and military autonomy and egalitarianism.

After all, the aggregate group of farmers, not urban savants, had kept the culture of the Greek city-state alive and began Western culture as we know it. And that very idea of a nation of self-employed, self-reliant individuals was also the founding principle of the American nation. Thomas Jefferson, in the romantic and often blinkered tradition of Western agrarianism, long ago lamented this intrinsic, unsolvable problem of family farming as the key to democracy: "I think our governments will remain virtuous for many centuries; as long as they are chiefly agricultural; and this will be as long as there shall be vacant lands in any part of America When they get piled up upon one another in large cities, as in Europe, they will become as corrupt as in Europe." An even greater tragedy, unforeseen by Jefferson, is that Europe has saved its farms far better than we. And so who is to be and not be corrupt they now know better than we.

Still, could not America at this late hour reverse that Hellenistic trend and relearn harder rural values that were with us at the beginning? The task is not, as in the case of the early Greeks, the physical challenge of scarce treasure, scant resources, poor soil, insufficient labor, or foreign invasion. We have the capital, the land, the people, the intelligence, and, unlike the Greeks, the historical precedents. But a rendezvous with simplicity, decentralization, and self-reliance entails starker psychological and ethical burdens where physical work would be respected, public service rewarded, and military duty required.

There is no hope that the family farm will return in the next century. Subsidy for corporate farming will only increase as technology, not agrarianism, is seen as the solution for the growing food crisis to come, for maintaining adequate stores at cheap prices. The thug and the financial pirate alike will not be reinvented as the local peach farmer or hardware store owner. But again could not we, one demands, at least adopt the *spirit* of the early Greeks and the contemporary farmer, at least transfer the values of the departed to post-*polis* America?

We could. But we will not, for we have not the courage. That renaissance involves an entire rejection of what we now hold so dear, a return to what we clearly despise: a common culture of peers, life away from the vortex of pelf and publicity, a firmness with the poor as well as the wealthy, an embrace of rural shame, a rejection of convenient urban guilt.

Could we really imagine a United States where public assistance for being unproductive is eliminated—for welfare mother and corporate irrigator alike, for the dank tenement and the tasteful vacation home, for the corner loiterer and the futures speculator? Could we envision Los Angeles and New York as failures of the human spirit, worthy of no dole, no handout, only to xerox an Alma or Lundburg as the closer ideal of the *polis*?

Do we want, through incentive and fine, a state where all citizens must be reeled in to vote on their local and national affairs, a community where the cargo of bankruptcy, divorce, and illegitimacy draws stern rebuke, not state sanction? Do we really desire an "intrusive government" that would tax annuity, interest, and dividend only to subsidize family-centered construction, manufacturing, and agriculture, a government that does not call stasis "stability" or rising employment "inflationary," that does not demand increased productivity and loss of good jobs at the same time?

And can the eighteen-year-old forgo his wasted semester of sociology at the local JC, the more eager Ivy League tyke sacrifice his first cog on the *cursus honorum* for a year of shared military service, where all in the American phalanx learn that they are, for all their various hues, residents of the same land, speakers of an identical tongue, fierce protectors by birth or choice of a long and distinguished heritage? Do we want a society where the ingestion of drug, the addiction to alcohol, and the conception of the bastard bring not solely empathy and counseling, but instead, as Rhys Burton once feared, also shame and contempt from a nearby community of tough-minded peers?

In that American *polis* one could not complain of big government while pocketing as relish his social security; could not desire to stand tall against foreign aggression without sending, like mustard-gassed Hans Henderson once sent Axel from his tiny farm, his only son into harm's way; could not talk of values only to strip the assets of his demented paterfamilias, parking him in a subsidized convalescent bed. He could not lecture on responsibility behind his gadgeted estate; could not call work his day's hedge, wager, and gamble; and could not equate living long with living well nor confuse uncreased skin and firmer flesh with probity and character. And so we would come to learn, as the Greek *geôrgos* and the American farmer have, the bitter lesson that total freedom is chaos, that what is legal is not always moral, that we are by birth, training, or character neither equal nor inherently good, that to vote, to work with one's hands, to defend the state, to raise a family, to pass on something better, to repair, not scavenge and cannibalize, the wreckage of neighbor and friend—to do all that and break no statute, undermine no covenant, is, after all, life's work, nothing more, nothing less.

II.

"But wait!" you counter, "all is not yet lost." As Americans, faithful to your optimistic culture and natural ethos, you demand a happy ending and you demand hope, the Greeks' "betrayer of reason." So can I not at least outline some concrete agrarian plan, some utopian scheme that might save the 1 percent that is lingering yet, perhaps even expand that 1 percent to 2 percent or 3 percent, who could serve as a needed counterpressure against all that takes us away from more sober democratic values?

There are at least three measures, I think, easily within our power that could arrest the further decline of the American yeoman, that could preserve indefinitely a strong minority as a check on the rest of us.

1. Eliminate the Department of Agriculture root and branch, without hesitation. Forget the Republican downsizing of government that proposes the loss of the Energy and Education departments. The Department of Agriculture is far more dangerous to the American health. Nor should we forget to prune away all state and county farm agencies as well. We must dare to go even further and cease all government regulation and "support" of agriculture other

than statutes governing food quality and safety. Stop all subsidized water, insurance, research, market orders, and production quotas. Whatever their claims of fostering rural development, whatever their Depression-era origins, they are now designed for and written to foster corporate agriculture.

The growth in size of the Department of Agriculture is commensurate with the decline of the family farmer; such is the nature of all similar government entities devoted to agricultural development, from Pharaoh's to Montezuma's. In our own environs, all during the 1980s the local University of California agricultural field station grew. It doubled its infrastructure of shiny new automobile and truck fleets; impressive new offices sprouted on its campuslike experimental farm—in the same decade as family farms right outside its vast fenced compound floundered and were lost. The State University at Fresno followed the same paradigm. Impressive new complexes devoted to viticulture and fruit production broke ground at precisely the same time grape and tree fruit prices were at an all-time low; this is no accident, but the logical coalescence of corporate, government, and university moneys. Under our present system, if we did not have Bart Refinos at the university, we would have to invent them.

Most agrarian reformers, I suppose, would disagree. They trust government over private business, instead of trusting neither. These would-be Gracchi assume that reformed bureaucracies might subsidize smaller growers, that an enlightened Department of Agriculture in the right nurturing hands—such as their own—could do noble things in encouraging more sustainable family-owned enterprises, that a university, field station, or college agricultural research team under their own prudent control might yet arrest the slaughter of family farmers.

Such thinking is naïve—if not dangerous—to agrarianism. Given the political reality of our present system of national and local campaign financing and the current practice of funding political activity in general, a benign government agricultural entity is now impossible. It is an iron law of American politics that corporate agriculturalists give to politicians and in return enjoy federal protection regardless of party, ideology, or purported creed. These are Roman estate owners, not Greek yeomen, their ground *latifundia*, not *klêroi* in the agrarian patchwork; and so it is naïve to believe that federal bureaucrats, local congressmen, and state legislators might use government to arrest their onslaught. Corporatization in California grew most rapidly, for example, under Democratic-controlled congresses and presidencies, which often bragged of idealism and professed faith in participatory government. As Democratic congressmen walked with Cesar Chavez, as they inflated the economy to favor production over investment, they voted simultaneously to bring federal water to corporate farms in violation of the national reclamation acts. It is no accident, then, that federal projects mostly are named for Democratic congressmen, for the largess of Democratic state and federal legislators who served only those who paid. Only with the elimination of the entire Department of Agriculture and the exit from agriculture cf its satrapies on the state and county level can a free market in American agriculture emerge.

For too long the debate between agrarians and agribusinessmen has raged over the economies of scale—whether megafarms are more efficient than family operations. It is now time to put the issue to rest by eliminating all direct and indirect cash grants to agriculture, which have historically gone almost always to the very largest of farms. Let the market have the final verdict. The result to family farms could not be worse than the events of the last half century. Peripheral benefits might accrue once the subsidy-giving umbilical cord has been cut: the American consumer might pay for the real cost of producing food; farm land might be banked and for a time go out of production; the despised yeoman might win real respect, not sympathy and misguided romanticism, as he beats back the top-heavy corporation.

2. Regulate commodity brokerage. Antitrust laws now apply to nearly all sectors of our economy except agriculture and baseball. Neurologists cannot easily own radiology labs where their own patients are sent to be scanned. Doctors are discouraged from controlling pharmacies where their prescriptions are filled. Foreigners are

not to buy television stations. Newspapers, radio stations, cable companies, and television interests are not to be controlled by one concern. So our economy is rife with government intrusion against monopolies that challenge and threaten the free market and serve to distort laws of supply and demand, profit and loss, through vertical control of supply, sale, and processing. There is precedent, then, for such legislation.

A bridle on the fruit broker is not antithetical to the practice of free market capitalism, not inconsistent with the elimination of government bureaucracies, but intended to correct monopolies and trusts which seek to distort and alter natural commercial transactions. Capital accrued in the merchandising of food could seek its expression elsewhere beyond the homestead. Only this way can we ensure that Ross Lee Ford will invest his sizable profits into a third or fourth Lexus, a condo on Maui, or a nicer tennis court rather than another farmer's one hundred acres of foreclosed orchard and vineyard.

Nothing in the last century has been more lethal to the American farmer than the accepted practice of processing and selling food. No other industry in America—not financial services, not used-car sales, not investment banking—has had such a sustained record of price gouging and outright theft, without regulation, without accountability or honesty to the farmers who make it all possible.

Those who sell and buy food—food distributors, food processors, food brokers—should not be able to own farmland. Such vertical enterprises encourage cheap commodity prices by owning or leasing vast acreages whose food can be produced at a loss on the understanding that greater profit will be recovered in its preparation and merchandising. Successful distributors and brokers, in moments of candor, will confess, as they have to me, that they do not mind crashes. The depression in commodity prices ensures food is grown at a loss that is not a loss for them, either short or long term.

Only in times of agrarian havoc can brokers dictate farm compensation; prices stay about the same for the oblivious consumer in the supermarket as the middlemen in between pocket still more of the farmer's percentage. Their own farming setbacks—ripe for tax losses and depreciation—become but small prices to pay for the greater profit of commission on and markup of the produce of others. Low commodity prices, of course, also depress land values and so encourage consolidation of farmland by the vertically integrated. The modus operandi is known to any in farming: Get produce cheap; have supply always dwarf demand; watch prices crash; use commission, processing charge, distribution fee, write-off, and depreciation to cancel out farm loss; and then buy land cheap. And that trend in their eyes is always positive. The fewer outside variables in the supply of food, the surer a golden age of agribusiness—the promised utopia when fifty or sixty thousand operations will grow not 66 percent, but raise, process, distribute, and sell *all* our nation's food. One does not have to be a conspiracy nut, only a mere historian, to see that the existence of family farms has been the only impediment to such a brave new world.

Legislation need not be complex. A Greek of the fifth century B.C. could chisel the statute into a stele in minutes: *No entity engaged in sending fresh or processed food across state boundaries can own, rent, or lease the farmland where it is grown.*

3. The two great evils that have ruined family farming—that, as the Greeks and the Romans knew, have always ruined family farming—are absentee ownership and vast size. Both could be outlawed, but that is illegal and unwise. Let a businessman own ten thousand acres of vines and trees and live seventy-five miles away in north Fresno. We do, after all, live in a free society and in a market economy where one's right to private property is a chief tenet of our American democratic legacy. But present-day legislation and tax codes are not written in stone; they have, after all, been flexible enough to evolve into support of *latifundia* and the ethic of the distant manor.

The federal reclamation laws—which provide water to corporations at below the market price—must be rewritten in the purest sense of the word: redrafted back to the spirit of their initial conception when they were devices used to settle and populate arid land with agrarian communities. Those who do not live on their own farms or own more than a prescribed size of agricultural land—somewhere between two hundred and two thousand acres, depending on the area, crops, and the role of irrigation—should not be eligible for federally subsidized water. Most pose as rugged ruralists; but, as I write, they are devising mechanisms for using their own subsidized corporate irrigation allotments as mere products to be sold to urban centers for cash. "We need our water," they cry, as they have you build and maintain their canals so that they may sell the water to Los Angeles and so prove that they are tough-minded water farmers, after all.

So property and irrigation taxes should be indexed based on size and the presence of residents, with the stiffest rates per acre given for the largest parcels without owner-occupied residence. Inheritance and estate taxes must be similarly calibrated to farm size and residence. We have had enterprise zones, duty-free centers, redevelopment incentives, and government-backed corporate loans; the use of tax code and investment statute to save a dying institution—the American family farm—is not without precedent and involves no bureaucracy or new entitlement. Again, the net effect of indexing taxes to farm size and residence would be to put more people back into farming, more farmers back into local communities, more profit and loss back in the hands of those who grow the food.

III.

Change of a different kind is on the horizon—not reform, but radical acceleration of a century-long evolution. And so I must end with my former and more pessimistic forecast. Indeed, rural futurists now talk of a welcomed third wave; they cheerfully envision genetic engineering, newly designed crop species to thwart the elements, frosts, insects, and idiopathic crop failure, the centuries-old banes of the wily agriculturalist. Generations of pragmatic expertise are now to be replaced by textbook formula and business acumen. Enriched water, scientifically blended fertilizers, and varieties of interactive herbicides are to be tubed intravenously through the soil to the "plant's" roots on precise, computerized schedules. Land and animals designed and then cloned for the machine are to be cultivated and harvested in large factorylike settings, run by an efficient cadre of elite engineers, economists, and responsive employees. The whole complex—and ultimately vulnerable—enterprise is to be capitalized through the profits of insurance and investment strategy, with financial gain its only motive.

The sad history of complex societies, ancient and modern, argues that bureaucracies grow, never shrink, and so suggests that these futurists—not agrarian romantics—have seen the real forecast over the horizon. Unproductive citizens multiply, rarely wane. Taxation, urbanization, and specialization are the harvests of elite legitimizing and nuancing classes—government, insurance, advertising, law, finance—who feed and clone from ritual, regulation, and regimentation. Family enterprises give way to corporate and governmental entities; in their wake mercenary enlistment, personal service, and public subsidy are offered to the landless and jobless detritus. Rural becomes urban, farmers suburbanites, as popular culture overwhelms agrarianism, never vice versa. Even language is refashioned to convince that work is drudgery, simplicity boredom, and charity entitlement, words themselves reinvented so that capital accumulation might become "success," chaos "diversity," selfishness "freedom," and the failed a "victim."

"Farm" size in America is to increase even more radically. Farm settlement will become ever more nucleated. Crops will be more specialized, labor more rare, and the total acreage of farmland in the United States always more reduced. All this will be packaged for the consumer through the promise of cheaper food and expanded urbanization onto prime farmland. It will all be debated, nuanced, but ultimately legitimized by the usual array of subsidized social and natural scientists.

Again, this description is no promised agribusiness utopia. Nor is this hysteria voiced by the shrill and paranoid with unkempt facial hair and buckskin tassels. It is a scenario unfolding now, right now, in our midst. Its next stages are proudly outlined in current agribusiness publications. Its ultimate evolution is under cheery discussion in university agriculture, biology, engineering, economics, sociology, and business departments. Farmhouses, sustainable agrarian communities, small plots, independent viable agricultural operations, crop diversity, children raised to toil on the farm—all that, of course, as we have seen this decade, will become rarer. Ultimately, it will be the stuff of pastoral literature, nostalgic film, and the occasional agrarian activist's call to action to resurrect an ideology that is to be no more.

About this growing American agribusiness of the Hellenistic and imperial stamp, much hostile is now being written. These needed publications are mostly worries over our present food supply. As calls to arms, they identify and assail the costly prerequisites of corporate farming: the environmental hazards of mass applications of synthetic chemicals into a fragile landscape, the eventual exhaustion of soil and water through nonsustainable agricultural practices, the specter of sudden agricultural catastrophe through fuel or fertilizer shortages, soil salinity, or mutated plant pathogens, and the growing pollution of the air and scarcity of water once farms are turned into housing tracts. They rightly point out that the loss of crop diversity and farm decentralization will eventually make these complex food systems—that is, all of us as well—as vulnerable to sudden collapse as the palatial agricultural economies of the pre-*polis* past. Finally, they believe, it will reach a point where it cannot go on any further, finally, as the scripture says, "the lord of the vineyard shall come and give the vineyard unto others."

But as I outlined in the preface and throughout this book, I am more concerned with the here and now, more worried, as Aristotle and Jefferson were, about the human cost of the empty countryside and the effect it has had on American society and our democratic institutions.

So 1983 and the culture of 1983 are not bleak agrarian nightmares to come but are now well over a decade old. As that world of family farming in California disappears beneath our feet, entombed beneath the asphalt of the grasping shiftless and oblivious cash-hungry, I wonder as the century, as the millennium ends, who will now be left uneasy with rain in September? Who will there be to prune until dark in fear of a Mohinder Kapoor? When the eastern valley is finally transformed, as it will be in a decade or two, into a single, hideous megalopolis, strung a hundred miles long beside the polluted freeway, who of these car-bound will fight the likes of a Vaughn Kaldarian for his ditch water to save his kin's salvation with five acres of early nectarines? Who will dismiss the lethal prognosis of the hurried oncologist and his entire cargo of aides, clerks, machines, and protocols with but a "Well there, Doc, I guess I've had it." I wonder, too, if a Rollin Buckler will step forward to tell the magnate, "Enough, no more"; if a Taro Kobashi will add his own, "No, that's not my way"; if a Rhys Burton will worry about any rumor about his kin in town.

When the agrarian parquet is finally overwhelmed, as it too must be by the suburban boxes, will there be any of the gray heads left to hobble out of the vineyard, to tell hardworking, nearly sunstroked young sons with their shovels, "Whatta day, boys. Let's house up now, it's as hot as a firecracker out there, and you've watered the vines"? Will there ever again be a khakied Philemon and a hairnetted Baucis on our creaking porch to warn you at ten years of age about the need for solemnity, obscurity, and piety—the sacred, the sole trinity needed to climb the hill and avoid the flood to come? And so will there be any of these, next century, left to stuff their children's brains with the inherited agrarian cargo, to teach them all the lore of trees and vines, and therein the age-old way—for so long the *only* way—to count and then discount the growing madness they see about them?

POSTCRIPT

Year's End 1994

But what has became of the people in the fields without dreams?

It was announced in the *Fresno Bee* recently that 1993 and 1994 were good farming years, that all the crops were harvested and most saved from rain. Those who lived these last two years know otherwise, know that it was a fitting, a typical end to the depression decade that began in 1983. Final 1993 raisin prices will be about $830 a ton—received in full only by January 1995—about 60 percent of the price we received eleven years ago before the crash. Fresh Thompson grapes not dried for raisins went to the wineries for $3 a sugar point, or about $60 a ton (the raisin equivalent of $300 a ton), about the same offer as twenty-five years ago. Some older farmers fearful of making raisins and without juice contracts picked their grapes only to have their trucks turned away at the winery gates. The half-informed muttered of European dumping on the international grape juice market. Half our persimmon crop rotted and moldered, left mummified on the trees still for want of a price to cover picking costs. That was far better than earlier years when it was picked, packed, and then thrown in the garbage, unpurchased after weeks in the cold storage. Four thousand boxes of plums sat unsold in Ross Lee Ford's icebox and were dumped for less than the price of the box and trucking, commission, and cold-storage fee. And there are murmurs already that a final settlement payment of $830 a ton for 1994 raisins is "way too high." We are back again to cutting off grape canes, the farmers' self-help program of a decade ago to take raisin vineyards out of production for a year or so.

The French proudly announced this winter that the EEC would not change agricultural policies, despite pressure from the participants in the GATT negotiations. A trade war is bluffed with the Japanese, who price our plums and pears with tariffs twenty times their worth. The Clinton administration just boasted that the inflation rate for 1995 should stay under 2 percent; the interest we pay this year on over $500,000 of outstanding loans to the local Production Credit Association fluctuates between 11 and 13 percent and is climbing monthly as I write.

After lengthy negotiations, Bus Barzagus's mountain ranch is apparently sold. After many false starts, a near-destitute blue-collar refugee from Fresno found financing and paid Barzagus about half of the money he had in it. Randy Messingale visited Alaska, looked for virgin land when the subdivision of his river ranch is complete. Barzagus himself, now retired, has lost a great deal of weight and does not eat red meat or smoke. He is to take my son fishing in the Sierra when the snow melts down a little. I want him to lecture Ax a little more about farming: He is now twelve and showing too much interest in tractors, disks, spray rigs, and the like, which my brother and cousin work on in our driveway. My father, Axel, is of no help; he still talks of the day when his preteen grandsons and nephews will be running the ranch with Logan and Rhys. Thank God that our preteen girls, Cat and Louisa Anna, have worked too many summers in the shed, have heard too many tales of Ross Lee Ford and Hugh Effin, and have enough sense never to want to farm—or to marry farmers.

I could not find a single Regal Red plum orchard left this year; a few Royal red seedless vineyards are up for sale in the paper. As I hinted, a few years after our catastrophe we replanted our old Regal and Royal ground with new nectarines and peaches; my late mother Catherine's life insurance policy had paid off some of the Regal and Royal losses, at least enough to obtain new financing for new plantings. Soon these plums, nectarines, and peaches produced beautifully, but they sold once again this August for two to four dollars a box, the identical price listed in my grandfather Rhys's 1953 diary entry under "1953 Tree-Fruit Prices Received." Ross Lee Ford thought the nectarines too big, the peaches too small, the nectarines too yellow, the peaches too red. Logan has just now bulldozed them out and is looking for capital to start again in that rich soil—phase III in the great tree and vine incarnation of our mortgaged ancestral Hendrickson farm near Lundburg.

Ross Lee Ford is finished with his new abode, a beautifully designed mansion on a gentle rise, with a commanding view over surrounding orchards and vineyards. An architect and landscape engineer must have quite tastefully blended together hill and house; it is a majestic, earth-toned temple that attests to the bounty of the orchards and vineyards of the San Joaquin Valley of California. A fenced artificial meadow and lake beneath the knoll lend a pastoral rather than purely agricultural flavor to the rural estate. Last Christmas season Ross admitted in a rare moment of candor that you can make money in agriculture if you don't grow food.

Taro Kobashi took my cousin, brother, and me to lunch; he is optimistic about the future of the raisin business but not so sold on the future of the "reformed and new improved" Valley Girl Raisin Cooperative.

ChiChi Martinez, without throne and kingdom, called out of the blue after years of silence. On his day off from driving a truck he wanted to bring his ten-year-old over to fish in our tule pond. I was flattered—no, honored really—remembering the files of his once-colossal labor phalanx and so felt terrible in disclosing to him that Willem Burton's oasis has been dry since 1983, the very year ChiChi could outgallop the IRS no more.

But Ulysses Ponce came; he thinks that he can now begin pruning in a day or so. Ponce also wants us to pack his newly acquired peaches and plums, but our line is too small to handle any fruit other than our own. He seems to think that the last four tree-fruit packers have stolen from him. Last year, after picking and delivering his plums to the packing shed, he said that he received a bill for $30,000 "for services rendered." I could not understand the packer's cryptic invoice and so was of no help. But it was at least clear that Mr. Ponce had paid the packer, broker, trucker, and consumer for the privilege of taking his fruit off the trees. Oh, he also denied the bothersome rumors of his own trouble with the IRS—Logan (and I think other farmers as well) have been asked by the Fresno IRS office to turn over any records of business with a "Mr. Ulysses Calderon Ponce, licensed California state labor contractor."

Not long ago Bus Barzagus, who visits on my days that I don't go into Fresno, inquired whether I had seen Rollin Buckler; he had recently bumped into Fallon Buckler, the Valley Girl field man and cousin of Rollin. I told him I had not seen Rollin Buckler in five years. But then oddly just a day later, out of the blue, Rollin Buckler himself drove into our yard in his father's newest pickup. "I tell you that was some lawyer we got back then, that Hightower, but I guess he's dead now," was about all he had to say. In an effort to revive our old crusade, I started to inquire about a persistent but completely unproven story concerning a Mafia connection to Larry Black. But he had a belly full of Larry Black and so had me quickly sign some documents putting a formal end to the Raisin Action Committee and drove off; as usual we parted friends until our next chance meeting in the next decade or so.

Just recently, as I was finishing this postscript, a ninety-mile-an-hour Ford 500 went airborne into the vineyard next to the house, tore out two hundred feet of Willem Burton's ninety-year-old vines, made it three and a half rows in from the paved road. Two Oaxacans scampered out and scurried away easily from our pathetic family posse, taking beer and keys but leaving their crushed clunker, our pureed vineyard stakes, wire, grapes, leaves, canes, and uprooted stumps. There are seven previous such car-induced wounds in the vineyards nearby, thousands of dollars of lost vines and production. I pass on the sermon from the poor overworked highway patrol officer; she'd seen four such phenomena already that Sunday afternoon.

The last bit of my father's retirement covered the 1993 farm loss of $60,000. Because of long-term debt and our unimpressive tax returns of the last decade, the production credit office threatens to cut us off after the 1994–1995 crop year; there are, after all, no more funds in anyone's bank account. We survivors and equity gorgers do not have generous life insurance policies and appear too healthy anyway. For the first time in 117 years, some of our land on my great-great-grandmother's place was rented out this year (to Hector Soares) to ensure that our vanishing borrowed capital can cover what land is left that we do farm. And some of Willem Burton's vineyard land even lies uprooted and unplanted—his original homesteaded parcel right above the dry tule pond.

For their annual Christmas dinner skit this past year, the kids updated *A Christmas Carol*, Scrooge being the local loan officer but tormented by two, not three, ghosts of decrepit yeomen—a past, a present, but no future specter. If only it was as simple as Cratchit and Scrooge, if only we knew we were the Cratchits.

A minimart is now being built in a persimmon orchard but a mile and a half away on the west side of town, and a sprawling HUD housing development is rising right next to it. The freeway to Fresno is now clogged and a brown haze engulfs Alma in the summer. The new RV dealer down the road also uprooted forty acres of vines, putting in gazebo, miniature kiddy train, and concrete pond; he claims that he, in Alma *not* Fresno, has the biggest American flag and helium blimp in the county. It's quite a complex on the nearby horizon.

Over my objections, when the 1994 loan was cleared, my brother paid all the children their full 1993 summer wages. But three weeks before he wrote the checks, my son asked me why I could not be more like my brother. Why indeed.